Elements of Ethics for Physical Scientists

Elements of Ethics for Physical Scientists

Sandra C. Greer

The MIT Press
Cambridge, Massachusetts
London, England

This book was set in ITC Stone Sans Std and ITC Stone Serif Std by Toppan Best-set Premedia Limited. Printed and bound in the United States of America.

Library of Congress Cataloging-in-Publication Data

Names: Greer, Sandra C., 1945-
Title: Elements of ethics for physical scientists / Sandra C. Greer.
Description: Cambridge, MA : The MIT Press, [2017] | Includes bibliographical
    references and index.
Identifiers: LCCN 2017007622 | ISBN 9780262036887 (hardcover : alk. paper)
Subjects: LCSH: Science--Moral and ethical aspects. |
    Scientists--Professional ethics.
Classification: LCC Q175.35 .G75 2017 | DDC 174/.95--dc23 LC record available at
https://lccn.loc.gov/2017007622

10 9 8 7 6 5 4 3 2 1

To my mother, Louise C. Thomason (1920–1998), who taught me that just because everybody is doing it does not make it right, and to my partner, Ruth E. Fassinger, who taught me that just because nobody is doing it does not make it wrong.

Contents

Preface

After all, the argument concerns no ordinary topic, but the way we ought to live.
—Plato, *Republic*[1]

Why do scientists need to learn about ethics? Scientists already have so much to learn: mathematics and physics, chemistry and biology and geology, computer technology and laboratory techniques, statistics. Then, within every subdiscipline, there is a vast literature to master. Why spend time on ethics?

We scientists teach and study ethics because we need to reflect on "the way we ought to live" in our professional lives. Some say that books such as this one are just advocating *good science*, that *good science* is learned in the course of one's education, and, thus, that no more need be said: it is self-evident. Good science—scientific research that is accurate and reproducible—is certainly a part of what we want to address here. You can do reproducible science, but fail to give appropriate credit to your predecessors or your collaborators. You can produce important studies, but misreport your data to enhance the credibility of your work. You can do significant studies and at the same time do irreparable harm to human or animal subjects.

We want to understand how to analyze and present results in honest and unbiased ways. We want to give proper credit to those who contribute, and not give credit to those who do not contribute. Because we do science with the financial support of the society in which we live, we need to consider how our work can be justified within our society. We teach and study ethics because we want to be good scientists: scientists who work with intellectual and personal integrity, and with compassion and care for the human race and for this planet.

There has been much attention in recent years to deliberate fraud in science—to falsification of data, plagiarism, and other behaviors that most us would immediately recognize as unethical. While this book addresses these misbehaviors, the main considerations are the daily issues facing us as working scientists, the everyday decisions that challenge our integrity as we study our data and deal with our colleagues. The goals of this book are to raise awareness of the many ethical dimensions of our work and our professional relationships, and to teach ways of considering and addressing ethical dilemmas. This book provides no prescriptions for perfection and gives no all-encompassing solutions, but instead strives to nurture our ethical intelligence, experience, imagination, and commitment, so that we know an ethical issue when we see one and so that we know how to think about it.

Why have a *book* about learning about ethics in science? Why not just use an online learning module, some of which are already available? This book and others like it are meant to provide the material for courses in which the background reading is done out of class and the class time is used for in-depth discussions and case studies. The time spent in discussions with several minds in play together leads to a change in how people think about solving ethical problems: how they recognize an ethical issue and how they analyze and respond to an ethical issue. That change in thinking is less likely to happen for a person interacting alone with a computer learning module.[2] Moreover, in real situations, ethical deliberations usually involve interactions among people.

Why do we need *another* book on ethics in science? *Elements of Ethics for Physical Scientists* differs from other books in that it addresses topics that are not always included in existing books. *Elements of Ethics* introduces both philosophy of ethics and philosophy of science. *Elements of Ethics* includes the issues of underrepresented groups in science, the use of science for weapons, and the place of science in public policy.

Elements of Ethics is also different in that it is aimed mainly at physical scientists (chemists, physicists), whereas most previous books have been aimed at medical and biological scientists. Chemists, physicists, and engineers can relate better to the content of the book if they find a familiar vocabulary, known personalities, and accessible case studies and discussion questions. *Elements of Ethics* does not include bioethics, but appendix D refers the reader to resources on bioethics. *Elements of Ethics* does not address issues of ethics in engineering practice, but is still appropriate for the study of ethics in engineering research.

An important part of training in ethics is the vigorous discussion of the fine points of ethical issues. *Elements of Ethics* includes a set of questions and case studies at the end of each chapter. The questions are divided into those that can be considered immediately and without further information (termed *discussion questions* and *case studies*), and those that require the reader to seek further information by consulting references or by an Internet search (termed *inquiry questions*). In each chapter or section of a chapter, there is also one case study, a *guided case study*, for which detailed questions are provided that lead the reader through the analysis of the case using the steps outlined in chapter 1 for ethical decision making.

Readers surely will find topics about which they want to learn more and can be guided by the further reading lists and notes for each chapter. Appendix B gives a general bibliography of books, films, and websites that relate to ethics in science.

Notes

1. Plato, *Republic*, trans. C. D. C. Reeve (Indianapolis: Hackett Publishing, 2004).

2. Brian Schrag, "Teaching Research Ethics: Can Web-Based Instruction Satisfy Appropriate Pedagogical Objectives?," *Science and Engineering Ethics* 11, no. 3 (2005): 347–366.

Acknowledgments

It is a pleasure to acknowledge the help of

Barbara Belmont on the history of lesbian, gay, bisexual, and transgender issues in the American Chemical Society;

Martin Benjamin on philosophy of ethics and on animal rights;

Gwendolyn T. Bush on data analysis and for other helpful comments;

Philip DeShong for sharing information and for reading draft chapters;

Ruth E. Fassinger on human participants in research and for comments on the entire manuscript;

William L. Greer on weapons research;

Donald T. Jacobs for helpful comments;

Marc A. Joseph on the philosophy of ethics;

Margaret A. Palmer for conversations about feminist views of science;

Arthur N. Popper and Robert J. Dooling for conversations about teaching ethics;

The Department of Chemical and Biomolecular Engineering at the University of Maryland College Park for supporting my teaching of a yearly course on ethics in science and engineering from 1995 to 2008; and

The Sonoma County Public Library, the University of Maryland College Park Libraries, and the F. W. Olin Library at Mills College for providing information resources.

Sandra C. Greer
Sonoma, California

1 What Is Ethics?

What is hateful to you, do not do to your neighbor: that is the entire Torah: the rest is commentary.
—Babylonian Talmud, Shabbat 31a

Ethics is about what is right and what is wrong in human behavior. The terms *ethics* and *morals* often are used interchangeably,[1] although a common distinction is that *ethics* refers to professional behavior and *morals* refers to personal behavior.[2] For the purposes of this book, we will not distinguish between the two terms because we will encounter many intersections between professional and personal behavior.

"From the dawn of philosophy, the question concerning ... the foundation of morality [or ethical behavior] ... has been accounted the main problem in speculative thought, has occupied the most gifted intellects, and divided them into sects and schools. ... And after more than two thousand years the same discussions continue, ... and neither thinkers nor mankind at large seem nearer to being unanimous on the subject, than when the youth Socrates listened to the old Protagoras."[3] Thus wrote John Stuart Mill in 1861, and matters remain the same to this day. It is worthwhile to try to understand the basic thinking of philosophers, theologians, and psychologists over these millennia, before turning to the question of right and wrong in the lives of scientists.

Essentials of the Philosophy of Ethics

There have been two main lines of thought in the philosophy of ethics.[4] Theory 1 assumes that the correct behavior is that which leads to the most happiness for all. Theory 2 asserts that each person should behave in a way that can be generalized as a way for everyone to behave. Other theories have been proposed, but these two theories have been and remain the most influential.

Theory 1: We should act so as to achieve the greatest good for the greatest number.

This first approach, known as *utilitarianism,* was begun by David Hume (1711–1776) and later developed by Jeremy Bentham (1748–1832) and John Stuart Mill (1806–1873).[5] Such an ethical framework, focusing on results rather than on process, is known also as *teleological* or *consequentialist.* The utilitarians propose that ethical decisions can be made from a balance sheet of all the good results and all the bad results that would follow from various possible actions. Good results and bad results are determined by the increase or decrease of total happiness, not just the happiness of the person acting. The action that results in the greatest overall good is the preferred action. If good results exceed bad results, then the end justifies the means. If bad results exceed good results, then the end does not justify the means.

A well-known example is the *Heinz dilemma.*[6] Assume that Heinz is a fellow who has a sick wife but has no money. Is it acceptable for Heinz to steal from the pharmacist the medicine that his wife needs in order to get well? How would the utilitarians advise Heinz? They would think about what behavior would result in the greatest good for the greatest number. In this case, Heinz and his wife would be very much happier if he stole the medicine and saved her life, while the pharmacist would be somewhat unhappy at the loss of his property. Utilitarians would claim that in this case, stealing achieves more happiness than unhappiness, that the end justifies the means, and thus that it is acceptable for Heinz to steal to get the medicine. Their philosophy would not motivate them to seek other solutions.

The problem with a utilitarian analysis is that the accounting of the total good and the total bad, short-term and long-term, that result from a particular decision can be very difficult to make. The original theorists thought of *good* as being pleasure and happiness, and *bad* as being pain and unhappiness. Pleasure and happiness come from physical, emotional, psychological, and intellectual satisfactions. Pain and unhappiness come as dissatisfactions from the same sources. But how does one tally completely all the sources of pleasure and all the sources of pain? There can be information that is lacking, and there can be unforeseen results. For Heinz's example, what if the medicine is very rare and hard to get, and the pharmacist had ordered it for the use of another patient? Then Heinz's theft would have an unanticipated bad effect for this other patient, an effect that was not included in the initial accounting of good versus bad. Moreover, a utilitarian balance sheet may allow very bad things to happen to some people (such as the other patient) so long as that is balanced by good things happening to other people.

Thus the theory that making ethical decisions by seeking an outcome of the greatest good for the greatest number is useful, but is not always practical, sufficient, or objective.

Theory 2: We should act only in ways that we would want everyone else to act.

The second approach is the famous *categorical imperative* of Immanuel Kant (1724–1804), who argued that there are universal rules that each person must follow, and that those rules are defined by their very universality.[7] Each of us should act only in ways that we would be willing for everyone to act, and then these acceptable behaviors reveal the general rules for behavior and provide a rational approach to moral decisions. This point of view, based on principles and intentions (or rules or values) rather than on outcomes, is termed *deontological*. For Kant, the paramount rule or principle for behavior was that human beings should never be used solely as a means to an end. Kant's view owed much to prior views of rule-based ethics, including the Ten Commandments, the Golden Rule, and the Buddhist Eightfold Path.

How would Kant approach Heinz's dilemma? Assume there are three universal rules in a Kantian system: (1) do not use people merely as a means; (2) do not lie; and (3) respect human life. Now, is it acceptable for Heinz to steal from the pharmacist the medicine that his wife needs in order to get well? The way the problem is posed, Heinz's only option is to steal in order to save a life, but that would violate the first rule, since he would be using the pharmacist merely as a means to an end. A better solution would be to seek an entirely different process, one that would avoid breaking any rules. For example, Heinz could borrow the money, or he could work it out with the pharmacist. A Kantian would be motivated to seek solutions that violate no rules.

Thus an ethical theory based on some set of rules can lead to a broader analysis of ethical problems than does the single rule of the greatest good for the greatest number. Of course, any rule-based system has its own problems. The choice of the *universal* rules may not be universally agreed upon. A conflict between rules may be unresolvable without requiring that one rule be ranked higher than another, and then that ranking may not find universal agreement. There are no perfect ethical systems.

An exploration of the full range of ethical theories is beyond the scope of this chapter, but the questions at the end of the chapter will introduce more of these other ideas.

An Ethical Value System for Scientists

It will be useful in addressing ethical issues in science to have an ethical system based on such a set of universal rules. There are a number of ways to set up such a system, and the simple one that follows can encompass the range of issues that scientists confront.

The rules for this proposed ethical system will be a list of *values*, where values are attributes that have high importance. For example, if *loyalty* is a value, then loyalty is seen as being very important. Philosophers of ethics have debated the nature of values, whether they are objective (the same for all people) or subjective (different for different people); absolute (the same in all cultures and times) or relative (changing with cultures and time); knowledge-based (having some basis in an external reality) or opinion-based (existing in the minds of individuals).[8] The general references at the end of the chapter can lead you to further reading and thinking about values.

This proposed ethical system for scientists aims for a simple set of values and begins with the setting of just two values: *life* and *truth*. Then the acceptance of these two values will be seen to imply that there are three other corollary values: *the universe, knowledge,* and *justice.* Many questions will remain about how to assign those values and how to implement decisions. For example, even if scientists can all agree to assign a high value to human life, disagreements will develop about what is human (embryonic stem cells?) and what is life (two-day embryos?). You are encouraged to think about this choice of a set of values, and perhaps to try to make your own revised value system.

1. We value human life.

Respect for human life can influence every aspect of science, from the questions we decide to investigate, to the plans we make for a research project, to the way we treat our colleagues, our predecessors, and others affected by our work. When we state that we value human life, we intend also to place value on individual human autonomy. *Autonomy* means that people have the freedom to optimize their own lives by making choices that suit their own individual talents, tastes, and circumstances.

2. We value truth.

Scientists seek to find the *truth* about the universe: what is consistent and verifiable. The idea of scientific truth will be considered at length in chapter 3. A value placed on truth implies that we must do science honestly, keeping to complete and accurate reports. Without truth and honesty, the enterprise of science could not proceed, and failures of truth and honesty hinder the progress of science.

Therefore *life* and *truth* form this basic value system for scientists. From these two values, three more values follow as corollaries:

3. We value the universe.

The universe in which we exist is essential to human life. It is logical that the universe itself be high on our list of values: without it, we are nothing. For scientists, this valuation supports the devotion of lives and resources to the study and understanding of the universe: it is the essential justification for science. Here value is assigned to the universe as a consequence of the value assigned to human life. We could certainly have started the other way around, with *the universe* as a primary value, and with *life* as a corollary.

4. We value knowledge.

The value placed on knowledge follows from the value placed on truth and the value placed on the universe. Scientific knowledge is the accumulation of truth about the universe. Because we value the universe, we value knowledge about the universe. Chapter 2 will analyze the nature of science, and chapter 3 will discuss how science gets done in ways that increase knowledge.

5. We value justice.

Justice derives from both *life* and *truth*. Justice in science requires respect for the lives of other people in science, and respect for the truth about their contributions. First, we want to be just to all those, past and present, who work with us in science; chapter 4 will explore the dimensions of justice within the community of scientists. Second, we seek to offer participation in science to all who wish to join us; chapter 5 will examine the inclusion of underrepresented groups in science. Third, we want to be fair to human and animal participants in research (chapter 5).

A Process for Ethical Decision Making

The five value statements discussed above constitute a value system: life, truth, the universe, knowledge, and justice. Ethical issues arise when all of these values cannot be satisfied at the same time, when there is a *conflict in values* or an *ethical dilemma*. Recall Heinz and his dilemma between saving the life of his wife and being just to the pharmacist. Psychologists and philosophers have proposed various ways of dealing with such conflicts.[9] For example, Kohlberg argued that the process is deductive and rational: Heinz values his wife's life, therefore he steals the medicine.[10] Gilligan believes that the process can be inductive and iterative, and can involve negotiation as much as reason:[11] Heinz considers analogous situations and develops as many other solutions as possible before making a decision that satisfies as many values as possible.

Whitbeck has compared ethical decision making to engineering design.[12] In both cases, (1) a problem is posed that has constraints on the solution; (2) there is no one correct solution, but rather many possible solutions within the constraints given; (3) there are, however, solutions that are clearly wrong, and "some solutions that are better than others."[13] The question for both engineering design and for ethical decision making is how to find the best of the possible solutions, meaning the solution or solutions that satisfy the most constraints. Whitbeck[14] and others[15] have proposed processes for finding the best solution.

A consideration of those ideas leads to the following seven-step ethical decision-making process that is more contingent and iterative than it is analytic and linear.

1. Name explicitly the values that are involved and consider those that are in conflict. For Heinz, he faces a compromise of value 1 above, human life (that of his wife), versus value 5, justice (fairness to the pharmacist). Sometimes you need to rank values, and sometimes you do not. Heinz may rank life above justice, but perhaps both can be satisfied.

2. Assess whether more information would be useful. For example, Heinz and his wife may seek a second medical opinion to be sure that she really needs the medicine. This step is, in fact, one of making sure that the statement of the problem is clear. The collection of more information will be a continuing part of the decision process.

3. List as many possible solutions as you can. For example, Heinz could list getting a loan, asking the pharmacist to permit payment over time, or sweeping floors for the pharmacist to make up the debt. It can be especially helpful at this stage to consult other people who have experienced similar problems, to see if they can help in generating more avenues for action.

4. The development of possible solutions often leads to the need for still more information. Heinz may need to review his health insurance policies to see if his wife's medicine is covered, or he may need to consult his employee credit union about the possibility of a loan, or he may want to talk to the pharmacist about his view of possible solutions.

5. Investigate these possible solutions in parallel, so that time is not lost if one solution proves unworkable. Heinz does not want to lose a lot of time planning on a loan if it turns out that he cannot get one; he wants to be working on the other avenues at the same time.

6. Be alert to changes in the status of the problem in the course of time that can lead to another iteration in the decision process. What if Heinz's wife gets better? Or worse? Or a new treatment appears?

7. Consider how each solution connects to the values that you identified. Decide on the best solution, and take action. This is far easier said than done! There may be cases for which different people reach different solutions. There may be times when no solution seems appealing, but still some action must be taken.

As stated, ethical theory and ethical systems have been argued over the ages. There is no consensus either about the structure for an ethical system, or about the choice of values or virtues or rules within a system, or about the resolution of any particular ethical dilemma using a given system. The value-based system used here is neither unique nor universal, but is just one way of framing an ethical system. It is a provisional and experimental. It may not always work as stated: some steps may not make sense in a given case, or may be taken in a different order. There may be times when a mixing of utilitarian ideas within this deontological system may be useful. Nonetheless, this system provides a starting point for the analysis of the ethical problems scientists face. The guided case studies at the end of each chapter or chapter section will provide more experience in thinking through ethical dilemmas in this way.

Summary: What Is Ethics?

A deontological ethical system for scientists is presented in which *life* and *truth* are taken as key values. These two values then imply three other values: *the universe*, *knowledge* of the universe, and *justice* for other people. Ethical dilemmas arise when all these values cannot be satisfied at the same time. The inevitable ethical dilemmas can be addressed by a process that seeks and considers as many solutions as possible before reaching a conclusion for action.

Guided Case Study on Ethics in Science: Cheating on an Exam

You are a student in the thermodynamics course that is required of all graduate students in chemistry. At midterm, the professor gives a take-home exam that counts for 33 percent of the course grade. Your best friend Emily is also in the class. Emily's mother has just been diagnosed with cancer and the prognosis is not good. Emily is spending a lot of time dealing with this problem and is unable to get her academic work done. Frantic and panicked, she asks you if she can copy your solutions to the thermodynamics take-home exam.

1. Name the values that are involved in this case and consider the conflicts among them. Are the values *life* and *truth* relevant? What about *justice*? How important is this exam compared to the stress in Emily's life? What does this portend for Emily's future as a scientist who cares about truth?

2.   Would more information be helpful? Can you talk more to Emily about this problem? Can you talk to another friend? To the professor? Is there professional counseling available for Emily at your university?

3.   List as many solutions as possible. Is there any argument to be made for letting Emily copy your exam? Could she ask for an extension of the deadline for the exam? Should she drop the course? Should she consider a leave of absence from the graduate program? Can you think of other solutions?

4.   Do any of these possible solutions require still more information? What is the effect on your friendship if you do not let Emily copy your exam? Will the professor allow Emily an extension of the deadline? What are the steps in dropping the course or seeking a leave of absence? What would be the consequences if she dropped the course or took a leave of absence?

5.   Think further about how you would implement these solutions. Can you help Emily to think through a timeline for her mother's illness and for her own future plans?

6.   Check on the status of the problem. Has anything changed? Is her mother better? Worse? Can you get a more detailed prognosis?

7.   Decide on a course of action. Can you help Emily to get counseling for her emotional turmoil and then to decide which course of action will work for her?

Discussion Questions on Ethics in Science

1.   Compare and contrast *legal* behavior and *ethical* behavior. What is an example of behavior that is legal but not ethical? Ethical but not legal? How do legal and ethical aspects overlap?

2.   Is there any connection between ethics and religion? Is a religious person invariably an ethical person? Is an ethical person necessarily a religious person? Can you think of historical examples of each case? Are some ethical principles shared or not shared among religions?

3.   What connection may exist between etiquette and ethics? Consider, for example, the practice of referencing previous publications by other people that form the basis for your own new publication. Is this etiquette or ethics or both? Can etiquette sometimes be founded in ethics?

4. How would you revise the set of five values suggested earlier for scientists? Would you choose different fundamental values? What would you add? What would you delete? How about loyalty? Integrity? Honor? Duty? Country? Money? Power? Happiness? Freedom? Tradition? Are these fundamental values, or are they corollaries to the two fundamental values of life and truth discussed previously? Or they can they be antithetical to truth and life?

Case Studies on Ethics in Science

1. You are a polymer chemist working for a company that produces plastic bags and containers. You read in the *New York Times* that plastic refuse is now a major contaminant in the world's oceans, disrupting ecological systems and threatening fish, birds, and marine mammals.[16] What does this mean to your company? To you as a citizen and as a scientist? What should you do? Analyze the problem using the seven-step ethical decision-making process set forth earlier.

2. You are a professor of chemistry and you are chatting with a colleague from the chemical industry at a meeting of the American Chemical Society. One of your graduate students is presenting a poster at the meeting and the colleague has met the student. The colleague says to you that she has an opening in her laboratory for a new chemist and that she plans to discuss this career opportunity with your student. You have been trying to persuade your student that she should aim for an academic career. How should you respond? Analyze the incident, again using the seven-step ethical decision-making process.

3. You are putting together your curriculum vitae in preparation for applying for your first job. You will have written six papers from your graduate and postdoctoral work. The first three papers have appeared in print, so you can list the exact references. Of the last three papers, one is submitted and under review, one is in a final draft on your computer, and the last one has not yet been started. You plan to list all of the last three papers as *submitted* on your curriculum vitae. Discuss in terms of the system of five values developed previously.

4. Recall an ethical issue that has arisen in your personal life and describe that issue. Analyze that issue using the seven-step ethical decision-making process.

5. Recall an ethical issue that has arisen in your professional life and describe that issue. Analyze that issue using the seven-step ethical decision-making process.

Inquiry Questions on Ethics in Science

1. Learn about the *logic trees* that are used in the discipline of logic to organize decision making. Construct a logic tree for the process for making ethical decisions. Can you apply that logic tree to one of this chapter's discussion questions or case studies?

2. The *social contract theory* of ethics proposes that an ethical framework arises from the mutual agreement of the members of a society, structured so as to provide for the benefit of each individual.[17] What are the issues in defining the members of this contract? For example, would animals be included? Would there be a need for ethics for Robinson Crusoe, a society of one? How would such an ethical contract be related to the legal system constructed by a society? How would such a contract relate to Henry Thoreau's ideas about civil disobedience?[18]

3. There are those who think that the only value needed for an ethical system is that of selfishness, and that the only motivation should be self-interest. This is the philosophy known as *ethical egoism*.[19] The essential idea is that everyone would be best off ("the greatest good to the greatest number") if each individual simply looked after his or her own best interests. What are the advantages and disadvantages of ethical egoism?

4. Aristotle and many others have thought that good ethical behavior arises from an innate capacity for personal virtue: a philosophy known as *virtue ethics*.[20] These virtues (including honesty and courage) are learned from the community. Then one operates from a set of virtues rather than from a list of values. What is the connection between virtues and values?

5. Carol Gilligan argues that boys and girls, men and women, think differently about ethical issues.[21] In her view, men focus on justice as a value and rules as a process, while women choose human connection as a value and negotiation as a process. What do you think? How might this have come to be true? How might men and women act differently in Heinz's dilemma?

6. Most professional associations have developed *codes of ethics* to guide their members in their professional lives.[22]

   a. Locate a copy of the code of ethics for your profession (e.g., chemistry, physics, chemical engineering).

   b. Consider the values exemplified in the code of ethics. Are they the same as or different from the values chosen for the ethical system in this chapter?

   c. Think of two examples of situations that exemplify sections of the code of ethics, and analyze those situations in terms of the seven-step ethical decision-making process.

7. Consider the ethical issues that can arise when an American company locates and operates a plant in a foreign country. What if the legal requirements for safety are less restrictive in this country than in the United States? What if the expectations for the treatment of women and minorities are not the same as in the United States? Look up and read about the 1984 Bhopal chemical plant disaster, then use this as an example of your points.

8. There are a number of nonprofit organizations (NPOs) and nongovernmental organizations (NGOs) that focus on the effects of human activities on the environment, including the problems of chemical manufacturing.[23] What are some of these organizations, and what career opportunities do they offer for scientists?

Further Reading on Ethics in Science

Brincat, Cynthia A., and Victoria S. Wike. *Morality and the Professional Life: Values at Work*. Upper Saddle River, NJ: Prentice-Hall, 2000.

Kidder, Rushworth M. *How Good People Make Tough Choices: Resolving the Dilemmas of Ethical Living*. New York: HarperCollins, 2003.

Pojman, Louis P. *Ethical Theory: Classical and Contemporary Readings*. 6th ed. New York: Wadsworth Publishing, 2010.

Rachels, James, and Stuart Rachels. *The Elements of Moral Philosophy*. 8th ed. New York: McGraw Hill, 2014.

Resnik, David B. *The Ethics of Science: An Introduction*. New York: Routledge, 1998.

Seebauer, Edmund G., and Robert L. Berry. *Fundamentals of Ethics for Scientists and Engineers*. New York: Oxford University Press, 2001.

Shrader-Frechette, Kristin. *Ethics of Scientific Research*. Lanham, MD: Rowman and Littlefield, 1994.

Weston, Anthony. *A 21st Century Ethical Toolbox*. 3rd ed. New York: Oxford University Press, 2012.

Notes

1. Caroline Whitbeck, *Ethics in Engineering Practice and Research*, 2nd ed. (Cambridge, UK: Cambridge University Press, 2014).

2. Edmund G. Seebauer and Robert L. Berry, *Fundamentals of Ethics for Scientists and Engineers* (New York: Oxford University Press, 2001).

3. John Stuart Mill, *Utilitarianism*, ed. Mortimer J. Adler, vol. 40, Great Books of the Western World (Chicago: Encyclopedia Britannica, 1990).

4. Louis P. Pojman, *Ethical Theory: Classical and Contemporary Readings*, 3rd ed. (New York: Wadsworth, 1998).

5. James Rachels and Stuart Rachels, *The Elements of Moral Philosophy*, 8th ed. (New York: McGraw-Hill, 2014).

6. Lawrence Kohlberg, "The Development of Modes of Thinking and Choices in Years 10 to 16" (PhD dissertation, University of Chicago, 1958); Lawrence Kohlberg, *Essays on Moral Development, Vol. I: The Philosophy of Moral Development* (San Francisco: Harper and Row, 1981).

7. Immanuel Kant, *General Introduction to the Metaphysics of Morals*, ed. Mortimer J. Adler, trans. W. Hastie, vol. 39, Great Books of the Western World (Chicago: Encyclopedia Britannica, 1990).

8. Harold H. Titus, *Living Issues in Philosophy: An Introductory Textbook* (New York: American Book, 1959).

9. Seebauer and Berry, *Fundamentals of Ethics*; James Rest, Muriel Babeau, and Joseph Volker, "An Overview of the Psychology of Morality," in *Moral Development: Advances in Research and Theory*, ed. James Rest (Westport, CT: Greenwood, 1986).

10. Kohlberg, "The Development of Modes of Thinking."

11. Carol Gilligan, "In a Different Voice: Women's Conceptions of Self and of Morality," *Harvard Educational Review* 47, no. 4 (1977): 481–517; Carol Gilligan, *In a Different Voice: Psychological Theory and Women's Development* (Cambridge, MA: Harvard University Press, 1982).

12. Caroline Whitbeck, "Ethics as Design—Doing Justice to Moral Problems," *Hastings Center Report* 26, no. 3 (1996): 9–16.

13. Whitbeck, *Ethics in Engineering*, 58.

14. Ibid., 61–68.

15. Eugene Schlossberger, *The Ethical Engineer* (Philadelphia: Temple University Press, 1993); Judith P. Swazey and Stephanie J. Bird, "Teaching and Learning Research Ethics," *Professional Ethics* 4, nos. 3–4 (1996): 155–178; Adil E. Shamoo and David B. Resnik, *Responsible Conduct of Research*, 3rd ed. (New York: Oxford University Press, 2015); Mike Martin and Roland Schinzinger, *Ethics in Engineering*, 4th ed. (New York: McGraw-Hill, 2004).

16. Charles J. Moore, "Choking the Oceans with Plastic," *New York Times*, August 26, 2014, A23.

17. Rachels and Rachels, *Elements of Moral Philosophy*; Thomas Hobbes, *Leviathan, or Matter, Form, and Power of a Commonwealth Ecclesiastical and Civil*, ed. Mortimer J. Adler, vol. 21, Great Books of the Western World (Chicago: Encyclopedia Britannica, 1990); John Locke, *Concerning Civil Government, Second Essay*, ed. Mortimer J. Adler, vol. 33, Great Books of the Western World (Chicago: Encyclopedia Britannica, 1990); Jean-Jacques Rousseau, *The Social Contract*, ed. Mortimer J. Adler, trans. G. D. H. Cole, vol. 35, Great Books of the Western World (Chicago: Encyclopedia Britannica, 1990).

18. Henry D. Thoreau, *"Walden" and "On the Duty of Civil Disobedience"* (New York: Holt, Rinehart, and Winston, [1849] 1997).

19. Rachels and Rachels, *Elements of Moral Philosophy*; Ayn Rand, *For the New Intellectual: The Philosophy of Ayn Rand* (New York: Random House, 1961).

20. Rachels and Rachels, *Elements of Moral Philosophy*; Seebauer and Berry, *Fundamentals of Ethics*.

21. Gilligan, *In a Different Voice*.

22. Jeffrey Kovac, "Ethics in Science: The Unique Consequences of Chemistry," *Accountability in Research* 22, no. 6 (2015): 312–329.

23. Lauren G. Heine, "Perspectives: Nonprofit Groups Come in Many Colors," *Chemical and Engineering News* 94, no. 39 (2016): 40–43.

2 What Is Science?

At heart ... science is about the telling of stories—stories that explain what the world is like, and how the world came to be as it is.
—M. Waldrup, *Complexity*[1]

The Nature of Science

Science grows out of our wonder at the world in which we live. Our curiosity leads us to ask questions about how things are made and how things work. We scientists collect information about the physical world, contemplate that information, and construct stories that organize the information into forms that we can comprehend and that will allow us to make predictions about the world.[2]

We gather information about the world by observation, when we examine the world as we find it, and by experiment, when we change something to see what happens. We refer to the information that we collect as *observations* or *data*. We refer to the stories that we devise to explain the data as *models* or *hypotheses* or *theories*. Sometimes the word *law* is used, but only when the model has become so well established that we cannot imagine it ever needing to be modified, which does not mean that it will never need to be modified!

In doing science, the first assumption is that there is an external world that is common to us all, and that we can obtain information about that world. Then *truth* in science means that the information gathered (observations, experiments, data) about the world and the models constructed (hypotheses, theories) are consistent, each with the other. This word *truth* is a complex one, and there are those who disapprove of its use on the grounds that we can never reach "The Truth," but many of us find truth to

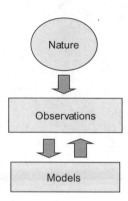

Figure 2.1
The nature of science.

be a useful word and even a beautiful ideal. Truth in science must develop over time, since data can be wrong and models can be wrong, and it is only over time that a stable consistency develops.

Science requires that we make observations and experiments on the external world. Chapter 3 will consider in detail the problems in making good observations. Once the observations are in hand, we can proceed to devise models by *inductive reasoning*: the generalization of particular observations to a general statement. The generalizations serve to collect and organize the data, to move from many special cases to the organizing principles that make for an efficient synthesis of all the data, and that allow predictions beyond the initial data. Without inductive reasoning, there would be only many special cases, no general models. Without inductive reasoning, there would be no science as we know it.

After we reach that model generalization, we can use deductive reasoning to make predictions from the generalizations to particular situations. *Deductive reasoning* is the logical application of a general statement to a particular case. We then test those predictions, revise the generalization if necessary, and repeat until we are satisfied that we cannot improve on either theory or experiments. When there is consistency between data and models, then *truth* is reached (at least tentatively).

We usually speak of the data as preceding the model. It happens just as often that the model precedes the data. For example, the Higgs boson was predicted from theory in 1964,[3] but was not confirmed by experiment until 2012.[4] Albert Einstein predicted from theory the existence of gravitational waves in 1916,[5] but the waves were not detected experimentally until 2015.[6] When the model precedes the data, that model does not just appear out of nothing, but is itself based in earlier data and earlier models. Indeed, even the data are based in earlier data and in earlier models, as will be discussed in chapter 3.

The term *scientific method* is often used for the process of collecting data and then finding a model consistent with the data, but this term is rarely used by working scientists. Indeed, the real interaction among models and experiments is convoluted, nonlinear, and iterative.

Problems with Inductive Reasoning

Inductive reasoning is intrinsically fraught with chances for error. Consider the following observations:

1. Toyotas use gasoline;
2. Hondas use gasoline;
3. Mitsubishis use gasoline.

Then the conclusion *by induction* is:

All cars use gasoline.

This generalization can be used to reason *by deduction* to predict that any other car (for example, a Ford) must also use gasoline. Three observations on cars have been generalized to make a statement about all cars.

Chalmers has listed three problems with inductive reasoning:

1. The number of observation statements forming the basis of a generalization must be large.
2. The observations must be repeated under a wide variety of conditions.
3. No acceptable observation statement should conflict with the derived universal law.[7]

First, there are only three observations in the car example. How many observations would be enough to support the generalization? Second, the observations were made only on cars made in Japan, so the observations were not made on a wide variety of conditions: maybe only cars made in Japan use gasoline. The sample of cars is *biased* in that it is not fully representative of all the cars that are manufactured. Third, only one exception to the generalization will suffice to invalidate it: as soon as an electric-powered car appears, the model is voided.

Thus inductive reasoning is daring and perilous, but dare we must, since we have no other way for science to proceed. An example of the success of inductive reasoning for *qualitative* data is the work of Charles Darwin. Darwin traveled widely and examined a vast number of animal and plant species before generalizing his observations into the theory of evolution, one of the grandest, most comprehensive scientific theories of all time. Such successes have happened in spite of the flaws in inductive reasoning.

Physical scientists are more likely to deal with *quantitative* data than with qualitative data. Statistical inference is a mathematical aid to inductive reasoning from quantitative data.

Statistical Inference for Quantitative Data

The first two of the three problems Chalmer listed for inductive reasoning have to do with how many observations are needed and how the conditions of the observations need to be varied, in order to justify a generalization. *Justifying* a generalization means being able to say that it is *probably* true.

Statistical analyses can be descriptive or inferential. *Descriptive statistics* just give a summary of data. For example, we can calculate the average temperature in San Francisco on August 1 over the last ten years; that calculation is a summary of a set of data about San Francisco. *Inferential statistics* tests a hypothesis by using a sample taken from a larger population in order to infer properties of that larger population. For example, we can measure the temperatures in ten cities around the world on August 1 each year for ten years, and use that information to test the hypothesis that the temperature of the earth is increasing over time. The sample is the set of temperatures in ten cities for ten years; the larger population is the temperature of the whole globe for all time.

Statistics deals with *random* errors, not *systematic errors*. In the example of measuring the temperature in San Francisco, a systematic error would result if the thermometers had been calibrated incorrectly and always read low by one degree. This *bias* or *offset* would make every measurement wrong (too low) by one degree. Statistics cannot correct for such a systematic error. A systematic error would also occur if we did not vary the conditions of the measurement. For example, the temperature near the San Francisco Bay may differ from the temperature inland, and this could cause a bias in the data.

On the other hand, if the experimenter read the thermometer some-times low by one degree and sometimes high by one degree and this error happened randomly, then statistics can deal with this random error. For quantitative data with random errors, inferential statistical analysis can give us a measure of the likelihood that the data fit a particular mathematical model.[8] For example, let model A predict that a particular parameter has a value of 0.33, and model B predict that this parameter has a value of 0.50. We measure that parameter on a number of samples and use statistics to calculate that it has a value of 0.33 ± 0.03, where ± *0.03* represents a 99 percent confidence interval. We are now 99 percent certain that model A is a better model for this phenomenon than is model B, since the value 0.50 is not within the 99 percent confidence interval.

Statistical analysis can determine how many measurements are needed in order to reach a particular level of confidence, and thus permit the design of a good experiment that avoids Chalmers' first problem of having a large enough number of observations. However, just as statistics alone cannot assure that there are no systematic measurements errors, statistics alone cannot assure that there is a broad and random set of samples on which to make the measurements: the second problem of "a wide variety of conditions" still remains. Moreover, statistics alone cannot tell us whether a particular model is correct, because more than one model may adequately describe the same data set. Nevertheless, an understanding of inferential statistics is vital in doing good science. There even may be occasions when a scientist will need the help of a professional statistician in order to design and analyze an experiment.

Falsification

The third in Chalmers' list of the problems with inductive reasoning has to do with what is called *falsification*. Philosopher Karl Popper argued that if any generalization must be consistent with all observations, then a single disparate observation is enough to invalidate that generalization.[9] The flawed generalization must be altered or totally replaced in the light of the inconsistent information. Any number of observations that agree with the theory will never be enough to prove the theory to be correct, because it is always possible that a new observation will disagree and prove the theory wrong. Popper maintained that this *falsifiability* is a neces-sary condition for a theory to be called *scientific*. It is not enough that the accumulation of evidence can support the theory; it has to be possible to disprove the theory.

However, for many philosophers of science, falsifiability is not an essential criterion.[10] First, when theory and experiments do not agree, it is not necessarily the theory that is wrong. Experiments are also fallible. Chapter 3 will explore that ways in which experiments can go wrong. Second, a theory that initially seems not falsifiable may become falsifiable as technology advances.[11] From the late-1600s, theory claimed that when a material body burns, it releases a substance called *phlogiston*.[12] This theory was overthrown after chemists learned how to collect, handle, and weigh gases. In 1774, Englishman Joseph Priestley discovered gaseous oxygen and showed that burning (combustion) uses up oxygen from the air. Soon thereafter, Antoine-Laurent Lavoisier in Paris showed that the weight of the oxygen used up from the air is added to the weight of the burned material.[13] The phlogiston theory had predicted that burning *reduces* the weight of the burning body, so that theory was falsified when Lavoisier found that a burning body *increases* in weight by the addition of oxygen. Last, even a theory that is found to be false can still be true enough under particular conditions to be useful. Quantum mechanics is "more true" than Newtonian mechanics, but Newtonian mechanics suffices for many practical purposes. Falsifiability may not be essential to science, but it is still useful in thinking about how science progresses.

Imagination in Inductive Reasoning

There is yet a fourth problem with inductive reasoning: a scientist must be able to imagine a generalization that will describe all the available data. The collection and analysis of the data involve the development of an intuition about the material—"a feeling for the organism"[14]—that is akin to the intuitions of artists and novelists. Deep involvement in the subject is surely the key to the creation of a generalization or a model from a set of observations. However, it is not at all obvious how the human mind makes such a synthesis. It is a magical step, a leap of the imagination, requiring not only full knowledge of the observations and their context, but insight, ingenuity, and creativity. Moreover, different scientists may come to different generalizations, depending on their individual *standpoints*, as will be discussed further in chapter 3.

Sometimes it has happened that two scientists have imagined the *same* new model at the same time. In 1925–1928, Werner Heisenberg and Erwin Schrödinger both conceived of quantum mechanics at about the same time, albeit in two different formats that were later determined to be equivalent.[15] It is also quite possible that, for the same data set, two different researchers will, by induction, arrive at two entirely *different* generalizations. The question of how the dinosaurs came to be extinct has been debated for decades; the same evidence has led paleontologists to two main theories—asteroid impact and volcanic eruption—to explain why all the dinosaurs died.[16] When two or more models describe the same data, then more predictions must be made from the models and more data collected in order to resolve the issue.

Likewise, once a model exists to describe a data set, it is still possible that other models will later appear that will also fit the data: just because a model agrees with the data does not make that model correct. For example, in 1913 Neils Bohr introduced a simple model of the atom as a nucleus with electrons circling the nucleus in fixed circular orbits.[17] The Bohr atom explained the frequencies of light emitted from hydrogen atoms, but about twelve years later this simple model was supplanted by the quantum mechanical models in which the electrons are not in fixed orbits, but have probabilities of being in particular spaces around the nucleus.[18] The theorists who developed quantum mechanics (Heisenberg and Schrödinger, in particular) had an even greater reach of imagination than did Bohr, and their model of the atom had even greater explanatory power than Bohr's simple model. Philosopher Thomas Kuhn referred to the doing of science within a given overall theoretical framework as *normal science* that is then disrupted by a *paradigm shift* when a new framework comes into acceptance.[19] An example of such a paradigm shift would be that from the Bohr atom to quantum physics.

Science as a Method

The scientific method of using observations to reach generalized theories is limited by the problems of inductive reasoning. The number of observations must be large and taken over a wide range of conditions. The observations can turn out to be wrong. It takes imagination and intuition to devise a generalization from the observations. Once there exists a generalization or model, it can never be proven to be absolutely true, but only can be shown to be consistent with all the available data. Any model can be challenged later by new data. Any model can be supplanted by a more successful model. Science proceeds by trial and error, collecting more and more data, modifying existing models, and sometimes abandoning an existing model or paradigm for an entirely new and different model. Indeed, the progress of science is analogous to the process of ethical decision making described in chapter 1: both proceed iteratively and asymptotically toward resolution of the problem posed.

The Community and the Careers of Scientists

Funding Scientific Research

Modern scientific research is a community enterprise that happens primarily in three settings: industry, government, and academia. Scientists in industry and government work with the financial support of their institutions, but scientists in colleges and universities receive little financial support from their institutions for their research. Instead, academic scientists must obtain grant or contract support from government agencies such as the National Science Foundation (NSF), the National Institutes of Health (NIH), the Environmental Protection Agency (EPA), the Department of Energy (DOE), and so on, or from private foundations such as the Petroleum Research Fund of the American Chemical Society (PRF), or from individual companies, such as E. I. du Pont de Nemours and Company. A *grant* provides support for research aimed at a particular scientific question, but with considerable freedom for the researcher in process and direction. A *contract* provides support but with more constraints, and requires that there be a specified product at the end of the contract. Grants and contracts include *direct* funds to the researcher, and *indirect* costs or *overhead* to the institution where the researcher works; indirect costs are negotiated to be 40–90 percent of the direct costs.

For government grants, the academic scientist must first submit a proposal (about fifteen pages of project description, plus many bureaucratic forms). In that proposal, the scientist must establish that the proposed work is scientifically important and has some broad relevance, that the proposed approach to the issue is workable, and that the scientist or *principal investigator* is competent to carry out the work plan. The competence of the scientist includes not only training and experience, but also access to laboratory facilities and the ability to recruit support personnel, including graduate students and postdoctoral associates. The proposal is then sent to several experts in the field for review and comment. The reviewers must approve the research plan overwhelmingly or the work will not be supported. Typically, government agencies reject about 80 percent of the proposals that they receive.

Publishing Research Results

Once funding is received and meaningful results obtained, the scientist must submit those results for publication in scientific journals. The research is not a part of the body of science until it is published and available for other scientists to examine and use. (Exceptions occur in industry, where results may be proprietary and not published in the open scientific literature.) Journals are published by nonprofit scientific organizations (American Association for the Advancement of Science, American Chemical Society, etc.) and by for-profit companies (Elsevier, Nature Publishing Group, etc.), but the process is the same for all. The scientist and collaborators prepare a detailed presentation of the research, and submit that manuscript to a journal editor.

The journal editor asks one to three experts (referees) in the field to review the manuscript. The experts must agree that the research is valuable and credible. *Valuable* means that the result is a contribution to the advancement of science. (The editor of *Physical Review Letters*, the premier physics journal in the world, used to say, "When an author has published a Letter on an interesting new effect in 'whifnium,' we do not want it followed later by Letters on the same effect in 'whafnium,' then 'whoofnium' and so on"[20]). *Credible* means that the work meets current standards for reproducibility, consistency, and verifiability. The assessment of credibility is difficult because referees are very unlikely to try to repeat experiments or to recalculate theories, but must rely on their own experiences as to whether the work is consistent with other published work and whether enough details are given to allow the work to be replicated. Replication and reproducibility will be discussed later in this chapter. Sometimes published journal articles turn out to be incorrect in spite of the referee process; retractions and corrections of erroneous papers will be addressed later in this chapter. The authors must also give proper credit to prior published research (see chapter 4).

In general, the journal editor will follow the recommendations of the reviewers, although the editor has the discretion to seek more reviewers or to act on his or her own opinion. There exists a host of journals, with a range of acceptance standards. There are now online journals that publish quickly, without referees. There are weak journals that assume names that are close to those of reputable journals. There are predatory journals that send out email invitations to submit, that are not dedicated to scientific progress, and that try to extract fees from those who submit; lists of such predatory journals can be found on the Internet. Good scientists want to publish their work in good journals.

One measure of journal quality is the *journal impact factor* (JIF), which is the total number of citations for articles published in that journal over a two-year period, divided by the total number of articles published in that journal, to obtain an average number of citations per article per year for that time period. It is then assumed that better papers get more citations, and that a journal with a larger JIF is a better journal. *Journal Citation Reports* (see website) gives the following JIF values for 2015: *Science*, 34.661; *Journal of Chemical Physics*, 2.894; *Journal of Organic Chemistry*, 4.785; *Physical Review Letters*, 7.645. These JIF numbers need to be used with discretion. Better papers do not necessarily get more citations: an erroneous paper can generate a lot of citations when others try to correct the work. The JIF values are averages and say nothing about the values of individual papers in the journal.[21] JIF numbers are certainly not useful to the four or five significant figures that are given: for example, the difference between 34.661 and 34.660 is meaningless in assessing journal quality.

Scientists belong to professional organizations (American Chemical Society, American Physical Society, etc.) that hold national meetings, at which scientists give lectures to one another about their research. The Gordon Research Conferences are interdisciplinary meetings with both invited lectures and contributed posters (see website). Many conference presentations are refereed in that proposed presentations are subjected to a review and approval process before the meeting. These meetings are important ways for scientists to disseminate their results and to meet other people with similar research interests.

The careers of scientists depend on their successful research efforts, as indicated by the quality and quantity of their published papers. For academic scientists, continuous and substantial grant funding has become the sine qua non of their careers. On the one hand, no grants means no money, no money means no research papers, and no research papers means no tenure, no promotions, no respect from colleagues. On the other hand, grants and papers lead to jobs and promotions, and to national and international recognition. But grants are a means to an end, and that the end is research papers published in refereed journals. Papers in refereed journals are permanent contributions to the body of science, and are the key legacy of a scientist.

Correction in Science

Scientists are human and humans make mistakes, but the structure of science provides for the eventual correction of errors.[22] There are three main modes for checks and corrections: review of proposals, review of manuscripts, and replication.[23]

Review of Proposals and Manuscripts

As discussed, scientific research is reviewed by experts in the field, both at the stage of a grant proposal and then at the stage of a manuscript submitted for publication. *Peer review* is the review of a grant proposal, where the adjective *peer* means that the proposal is reviewed by four to six colleagues who work in the same research area, but these reviewers may be not be peers in that they may be more senior, more distinguished, more experienced. Indeed, senior scientists are more likely to be selected as reviewers than are junior scientists. Peer review asks whether the proposed work will make an intellectual contribution to the field, whether the proposer is competent to do the work, and whether the proposed research will make any broad contribution to science or to society or both. This review is a way of making sure that the research project has good prospects before funds are invested in it. The *referee system* is the review of scientific research papers before publication in journals, again by several experts in the field. This review aims to determine whether the research in the paper is valuable and reliable, and whether the presentation is understandable to other experts. The term *reviewer* will be used here for both review of proposals and refereeing of papers.

There are issues that can complicate the reviewing of proposals and papers. For example, what if a reviewer is a good friend of the author and is therefore predisposed to give a good review? Or what if the reviewer is a competitor of the author and inclined to be unfairly critical of the work? The National Science Foundation addresses such *conflicts of interest* (see chapter 6) by requiring each proposer to list his or her collaborators over the previous five years, including graduate and postgraduate mentors, and graduate and postdoctoral mentees. A more difficult issue in reviewing papers and proposals is the possibility of *implicit* bias. It is well established that all of us have biases based on age, race, gender, and ethnicity, as will be considered further in chapter 5.[24] There can also be biases based on institutional status (an Ivy League university versus a Third World university) and personal renown (a Nobel Prize winner versus a new assistant professor).

In the physical sciences, the common practice in reviewing has been *single blind* review, in which the reviewers or referees know the identities of the authors, but the authors do not know the names of the reviewers. Single blind review protects the reviewers from retribution from the authors, but does not protect the authors from either conflicts of interest or implicit bias on the part of the reviewers. Moreover, reviewers can inadvertently reveal their identities to authors by the nature of their comments, such as requests that their own papers be cited. In the social sciences, review is more often *double blind* in that authors do not know the identities of reviewers and reviewers do not know the identities of authors. Double blind review does protect the authors from bias and conflict of interest, but is difficult because authors may need to reference their own earlier work and thus reveal their identities. *Triple blind* review means that even the journal editors do not know the identities of authors (although the editors need to know the names of reviewers because they assign them), since editors are also susceptible to bias and conflicts of interest. There may be more use of double blind or triple blind review in the physical sciences in the future. In the end, the program officer for a proposal and the editor for a journal must weigh the comments of the reviewers and decide whether to fund the proposal or publish the article. Other ethical issues for reviewers and referees, such as confidentiality, are discussed in chapter 4.

There are changes under way in scientific publication practices that stem from the development of the Internet. First, there are efforts to reconsider the referee process. Thirty years ago, the referee process depended on the submission of hardcopy reviews from referees to editors. Now all communication is electronic. Manuscripts can be made available online for comments and discussion (anonymous or signed) from the entire community, in place of or in addition to the usual referees. *Atmospheric Chemistry and Physics*, an online journal published by the European Geosciences Union, has such an *open review* process. It is possible to publish online all the reviews and the responses of the authors, and it is also possible to hold an online discussion after the formal publication of the paper. Thus there is an exciting prospect of much more exchange within the community, before and after the publication of a paper.

The other area of change in scientific publication is the result of pressures from both the scientific community and from the public to make the results of research available more quickly and more broadly. First, the process from the submission of a manuscript to a conventional refereed journal to its publication can take six to nine months. This is a long time for other researchers to wait for information that is critical for their own work. Scientists are beginning to make *preprints* (new manuscripts that have not been refereed and have not been published) available on Internet websites such the arXiv.org website for physics, astronomy, mathematics, and computer science, and the ChemRxiv.org website of the American Chemical Society.[25] Some journals accept submissions that have appeared on preprint servers, but some reject such papers and consider preprints to count as true publications. Unpublished manuscripts are also available under an open review process.

Second, researchers without easy access to scientific journals as well as some members of Congress argue that when scientific research is paid for by public funds, that research should be available to the public without charge.[26] *Open access* refers to the free and open distribution of published scientific research over the Internet. The highly-rated refereed journals tend to have high subscription costs (about $30,000 a year for a library) and high costs for the downloading of individual papers (about $30 each) that make the scientific literature inaccessible for many scientists and for the public. The full solution of this problem is not yet clear, since the journal publishers need financial support and resist free access to their publications. One approach is the development of rapid publication and open access (but still refereed) journals such as *Public Library of Science* (*PLoS*). The American Chemical Society started the open access journals *ACS Omega* in 2016 and *ACS Central Science* in 2015. The journal *Science* started the open access journal *Science Advances* in 2015. Another approach is the recent requirement by a number of funding entities that publications be made publicly available, perhaps after some specified period of time. Already, publications resulting from research supported by the National Institutes of Health must be made available through the NIH PubMed Central (PMC) online library.

Thus, investigations are reviewed before they begin, the results are reviewed before they are published, and there may soon be online reviews after publication. These reviews check the correctness of scientific results as they are produced, but reviews alone do not assure that all published work is correct. Science depends, in addition, on the long-term replication and reproducibility of that work by other scientists.

Replication and Reproducibility

The term *replication* means the exact duplication of a published study, and the term *reproducibility* means consistency of the published study with related work under other conditions.[27] Replication and reproducibility are essential to the growth and development of science, but do not always happen, for several reasons.

1. The credit and glory in science go to the scientist who first publishes a new result. There is little glory for repeating the work, and thus not much motivation.
2. Journals and the referees making decisions for journals have not been very interested in publishing replications. However, recent public discussion of the importance of replication has led to changes in journal policies so as to encourage papers that replicate earlier work.[28]
3. Scientific research is expensive in time, money, and materials. It is difficult to obtain money from granting agencies just to replicate an existing study.

The motivations that do lead to repeating a scientific investigation are either (1) to learn a technique, or (2) to prove the original work to be wrong. A researcher may repeat a published investigation if he or she is planning to work in the same area and wants to check his or her techniques against a known and trusted result before moving on to new investigations. Or a researcher may have some reason to distrust a published result, and thus set out to repeat the study.[29] When a scientist sets out to replicate a study, it is not easy. Often the published paper will not contain enough details about the research for a valid and complete replication to be possible. Or the research may require special experimental expertise that takes long practice to master. Sometimes the sample studied is one that is hard to obtain (e.g., a rare event in a cloud chamber, or a group of people who are no longer living). Often the original data are not available for a good comparison: they may have been published in a reduced, averaged form or only as a graph, and not as tabulated, primary data (see chapter 3).

Many journals have addressed the replication problem by increasing requirements for experimental and statistical details and by requiring the submission of complete data sets for electronic storage. The journal *Science* has added a board of statisticians to review papers where the statistical analysis is critical.[30] There is also a movement to make more data sets available directly and freely by depositing the data in repositories, managed by information experts.[31] The collected data then allow easier access for replication and reproducibility checks, but they can also allow *meta-analyses* of the data: other scientists can use the stored data sets to explore new questions.

Because replication is so difficult, scientific results are more likely to be tested by reproducibility (consistency with related results) than by replication (exact duplication). If the original study starts to look anomalous in the context of later, related work, then the original work will either be replicated or (more likely) just discounted.

Honest Error

Honest mistakes do happen in science.[32] It is good for science that scientists be able to take chances in their work, make mistakes, and move on without penalty. Sometimes errors in science are realized by the same scientists who made them. Journals have a mechanism, a short article called an *erratum*, by which a scientist can publish a correction of an error in his or her own paper.

If a paper is found to be entirely wrong, then the author can voluntarily *retract* the paper, meaning it is withdrawn after it has been published. Papers can be retracted by the journal editors in cases of serious errors that invalidate the paper and are not acknowledged by the authors, or in cases of fraud (see chapter 3). It is a service to the community and to science to correct one's own mistakes, since this keeps the literature cleared of error and can save other people time and money. There is some indication that the journals with the highest impact factors (JIF, see earlier discussion) also have the highest retraction rates, but this is not well established.[33] If this is true, it could be because high-impact journals publish more adventurous research, and because scientists rush into print in high-impact journals for the associated prestige, perhaps not vetting their work completely.

An example of the success of the internal correction system of science is the case of *polywater*.[34] In 1962, Russian scientist Nikolai Fedyakin published an article claiming that ordinary water changes its structure when condensed into glass or quartz capillary tubes.[35] This work on *modified water* or *anomalous water* or (later) polywater was taken up by a better-known Russian scientist, Boris V. Deryaguin (or Deryagin), who produced a series of papers with various colleagues (including Fedyakin). The modified water was found to have very different properties from normal water: more viscous, denser, higher boiling point, lower freezing point. In 1966, Deryaguin presented his work at a meeting of the British Faraday Society, where he claimed that solid surfaces could permanently change liquid structure, that this could happen to liquids other than water, and that this new form of water would be the stable form into which all water would ultimately transform.[36] Over the next two years, scientists in Europe and the United States became aware of the claims about polywater, and became interested in the impact of polywater on such processes as cloud formation, plant capillary action, and physiology. The U.S. Office of Naval Research began to fund polywater research. The mass media started reporting on polywater because of the threat that all water could turn into polywater. Theorists began to propose various polymeric configurations of water molecules that would explain the experimental results. More and more scientists became involved until some two hundred people were working on polywater. Publications and discussions continued until 1970, when Denis Rousseau and Sergio Porto[37] of Bell Laboratories (and others[38]) proved that the anomalous water was basically dirty water, containing impurities leached from the capillary walls. The saga of polywater was over by 1974.

It can be argued that scientists were too quick to believe that polywater could exist and were not critical enough of the experiments. However, there were never issues of deceit or data fabrication (such misconduct will be discussed in chapter 3). The scientists were working with integrity and (with some exceptions) with cooperation and courtesy. In the end, the collective process of science worked and polywater was found not to exist, even if at a significant cost in money and effort. There was also some cost to the reputations of the polywater advocates: while scientists respect those willing to admit their mistakes, mistakes that are too numerous or too significant can affect professional reputations.

Summary: What Is Science?

We scientists have an ethical obligation to seek the truth about the universe in which we live. We pose questions that lead us to models or to observations. We generalize observations into models. We make further observations to test the models. We revise the models to accommodate new observations.

We can miss the truth at any step. We can fail to ask the right questions. Our observations are flawed. Our inductive reasoning is suspect. Our models are tentative. Referees and reviewers can fail to find errors. Replication does not necessarily occur. How do we ever get any good science done? Chapter 3 will explore how we can try to minimize our failures.

Guided Case Study on the Nature of Science: Failure to Replicate

You, a new assistant professor, have attempted to replicate an experiment published by a group led by a very distinguished scientist. Your replication effort has failed: you do not get the same results that they got. What should you do next?

1. Name the values that are involved and consider the conflicts among them. Review chapter 1 and think about the set of values for ethical decision making.
2. Would more information be helpful? Whom can you consult? Researchers at your university? The original authors? Can you use Web of Science or another database to determine whether other researchers have cited the original paper and what their comments have been? Do you need to repeat your own work? Should you have someone else review your notebook, or even repeat your procedures?
3. Assuming that you conclude that you are correct in your experimental result, list as many solutions as possible. Are the original authors open to collaboration on a corrected paper? Can you spend time in their

laboratory to learn their techniques and see how they differ from your own? If they are not cooperative, then will you submit a paper based on your own work, for which the original authors are likely to be chosen as referees? Or should you ask that they not be referees?

4. Do any of these possible solutions require still more information? What if some of the original authors want to cooperate and some do not?

5. Think further about how you would implement these solutions. If the original authors are not cooperative, do you risk that their antagonism will have a negative effect on your career? What if they are consulted when you are reviewed for tenure?

6. Check on the status of the problem. Has anything changed? Have any new papers been published that affect your thinking?

7. Decide on a course of action. What if you write a paper on your own and the journal rejects it, based on the reports of the referees?

Discussion Questions on the Nature of Science

1. Is computer science a science? Economics? How do we identify a science? Do we need to specify what is or is not science?

2. Is there such a thing as the *scientific method*?

3. If science is the study of an external world that is common to all of us, then how do we think about the internal worlds that are not common to all of us? How do we expand the definition of science to include psychology?

4. What problems may occur in double blind reviewing of papers and proposals? On the one hand, would the issues in double blind refereeing for papers be different from the issues for proposals? On the other hand, what if the process is completely open and the identities of both authors and reviewers are known to the other?[39] List the advantages and disadvantages of single blind, double blind, and open reviews.

5. Computer simulations of physical, biological, and social phenomena have become a major part of scientific research. Are computer simulations theories, or are they experiments, or are they something else? How do they fit into the scheme of science?

Case Studies on the Nature of Science

1. Professor Smith is a young organic chemist, just starting her career. There is much pressure on her to succeed, since she is the first female organic chemist in her department and many have high hopes for her. She has undertaken the synthesis of an anti-cancer agent as her major research effort. Five years into the project (and one year from her tenure review), she is stuck at a final and crucial point in the synthesis. Nothing seems to work. Then she receives by email a proposal to review for the National Science Foundation. The proposal contains information that will solve her problem. If she uses that information, she will help to cure cancer, get tenure, and live happily ever after. However, information in proposals is considered confidential and not to be used to the advantage of the reviewer. What should she do?

2. Dr. James is a distinguished senior physics professor at a very prestigious university. His days are very full, with teaching and research and service. Two papers in his field of fluid mechanics lie on his desk for his review. Paper A is by an established scientist with a long career, many papers, and many awards (including membership in the prestigious National Academy of Sciences). Paper B is from a newly appointed assistant professor, working at a minor university. James knows Author A, but is not a collaborator and has no formal conflict of interest in reviewing that paper. James does not know Author B, and, in fact, has never even heard of this author. What unconscious biases may affect James's review of these papers? Would status matter? What if one of the authors is clearly female?

3. Jack Stone, an industrial chemist, realizes that a paper that he coauthored ten years ago as a graduate student is not correct.

    a. What if the error is a misplaced decimal point in a table of data? What should Jack do?

    b. What if the data are correct, but the interpretation of the data is not correct? What should Jack do?

    c. What if the data are not correct, but are completely wrong? What should Jack do?

    d. In each case a, b, and c, what are the roles of Jack's coauthors?

Inquiry Questions on the Nature of Science

1. Some philosophers argue that science depends on the time and cul-
    ture in which it develops and thus that science is *socially constructed*.[40]
    In particular, feminist philosophers argue that science as developed
    by men retains a gendered imprint that can be expected to change as
    more women participate (see chapter 5).[41] Read more about this point
    of view. What is an example of the effect of social context on scientific
    thinking? Is this a more relevant criticism in social and biological sci-
    ences than in physical sciences? How does this view fit with the view of
    science given in this chapter? Chapter 3 will revisit how the *standpoint*
    of the scientist affects how science gets done.

2. Failures of replication have been of much concern in social and life
    sciences. In psychological research, "only 39 percent [of studies] could
    be replicated unambiguously."[42] Read more about this issue. Do retrac-
    tions occur more in some disciplines than in other disciplines, and if
    so, why? Are the data on retractions sufficient to make statistically firm
    conclusions? Are there lessons here for physical scientists?

3. The impact factor of a scientific journal (JIF) was discussed in this chap-
    ter. What is the *h-index* of an individual scientist? How can each of
    these factors affect the career of a scientist?

4. Look up the website of the Meta-Research Innovation Center at Stan-
    ford University, METRICS. This center was established to study how
    research gets done and how reproducibility is achieved. What are the
    center's methods? What disciplines are its focus?

5. Sometimes a theory predicts a phenomenon, then experimentalists
    try to measure the phenomenon, and they can find no evidence of
    the predicted effect. Reports of such *negative results* have not been wel-
    come in the scientific literature, but they represent real information
    that can help other investigators, as information about paths that do
    not work.[43] An example would be the attempts of Joseph Weber at the
    University of Maryland, College Park to detect the gravity waves pre-
    dicted by Einstein. Read about attempts to include negative results in
    the body of knowledge. What was the effect of Weber's work on later,
    successful experiments on gravity waves?

6. Philosopher Thomas Kuhn proposed that normally science proceeds by
    the accumulation of data and the adjustment of theory within a given

paradigm, where that paradigm is the reigning theoretical framework.[44] Then there can occur a change of the paradigm by the introduction of a revolutionary new theory that offers a more coherent and more complete explanation. An example of this change of paradigm is the shift from Newtonian physics to quantum physics, where quantum physics includes and extends Newtonian physics. Read more about Kuhn's influential thinking. How does this model fit with this chapter's discussion of how science works?

Further Reading on the Nature of Science

Baldwin, Melinda. "In Referees We Trust." *Physics Today* 70, no. 2 (February 2017): 44–49.

Bloomfield, Victor A., and Esam E. El-Fakahany. *The Chicago Guide to Your Career in Science: A Toolkit for Students and Postdocs.* Chicago: University of Chicago Press, 2008.

Briggle, Adam, and Carl Mitcham. *Ethics and Science: An Introduction.* Cambridge, UK: Cambridge University Press, 2012.

Bronowski, J. *Science and Human Values.* New York: Harper and Row, 1965.

Chalmers, Alan F. *What Is This Thing Called Science?* 4th ed. Cambridge, MA: Hackett Publishing, 2013.

Committee on Science, Engineering, and Public Policy. *On Being a Scientist: Responsible Conduct in Research.* 3rd ed. Washington, DC: National Academy of Sciences, 2009.

Derry, Gregory N. *What Science Is and How It Works.* Princeton: Princeton University Press, 1999.

Firestein, Stuart. *Ignorance: How It Drives Science.* New York: Oxford University Press, 2012.

Firestein, Stuart. *Failure: Why Science Is So Successful.* New York: Oxford University Press, 2015.

Franklin, Allen. *The Neglect of Experiment.* Cambridge, UK: Cambridge University Press, 1986.

Franklin, Allen. *Can That Be Right?* Boston: Kluwer Academic, 1999.

Galison, Peter. *How Experiments End.* Chicago: University of Chicago Press, 1987.

Giere, Ronald N. *Understanding Scientific Reasoning.* 3rd ed. Chicago: Holt, Rinehart, and Winston, 1991.

Gratzer, Walter. *The Undergrowth of Science: Delusion, Self-Deception and Human Frailty*. New York: Oxford University Press, 2000.

Hacking, Ian. *The Social Construction of What?* Cambridge, MA: Harvard University Press, 2000.

Johansson, Lars-Göran. *Philosophy of Science for Scientists*. New York: Springer, 2016.

Kovac, Jeffrey. "Ethics in Science: The Unique Consequences of Chemistry." *Accountability in Research* 22, no. 6 (2015): 312–329.

Livio, Mario. *Brilliant Blunders from Darwin to Einstein: Colossal Mistakes by Great Scientists That Changed Our Understanding of Life and the Universe*. New York: Simon and Schuster, 2013.

Macrina, Francis L. *Scientific Integrity: Text and Cases in Responsible Conduct of Research*. 4th ed. Washington, DC: American Society for Microbiology Press, 2014.

Medawar, Peter B. *Advice to a Young Scientist*. New York: Harper and Row, 1979.

Rosei, Federico, and Tudor Wyatt Johnston. *Survival Skills for Scientists*. London: Imperial College Press, 2006.

Rothenberg, Albert. *Flight from Wonder: An Investigation of Scientific Creativity*. New York: Oxford University Press, 2015.

Sindermann, Carl J. *Winning the Games Scientist Play*. New York: Plenum Press, 1982.

Sindermann, Carl J. *The Joy of Science: Its Excellence and Its Rewards*. New York: Plenum Press, 1985.

Traweek, Sharon. *Beamtimes and Lifetimes: The World of High Energy Physicists*. Cambridge, MA: Harvard University Press, 1992.

Weinberg, Steven. *To Explain the World: The Discovery of Modern Science*. New York: Harper, 2015.

Wolpert, Lewis. *The Unnatural Nature of Science: Why Science Does Not Make (Common) Sense*. Cambridge, MA: Harvard University Press, 1992.

Wolpert, Lewis, and Allison Richards. *A Passion for Science: Renowned Scientists Offer Personal Portraits of Their Lives in Science*. New York: Oxford University Press, 1988.

Wolpert, Lewis, and Allison Richards. *Passionate Minds: The Inner World of Scientists*. New York: Oxford University Press, 1998.

Woodward, James, and David Goodstein. "Conduct, Misconduct, and the Structure of Science." American Scientist 84, no. 5 (1996): 479–490.

Wooton, David. *The Invention of Science: A New History of the Scientific Revolution*. New York: Harper, 2015.

Notes

1. M. Mitchell Waldrup, *Complexity: The Emerging Science at the Edge of Order and Chaos* (New York: Simon and Schuster, 1992).

2. Sandra C. Greer, "Truth and Justice in Science," *The Hexagon of Alpha Chi Sigma* 87, no. 4 (Winter 1996): 67–69.

3. Peter Higgs, "Broken Symmetries and the Masses of Gauge Bosons," *Physical Review Letters* 13, no. 16 (1964): 508–509.

4. Adrian Cho, "Higgs Boson Makes Its Debut after Decades-Long Search," *Science* 337, no. 6091 (2012): 141–143.

5. A. Einstein, "Näherungsweise Integration der Feldgleichungen der Gravitation," *Königlich Preußische Akademie der Wissenschaften zu Berlin. Sitzungsberichte*, June 22, 1916, 688–696; A. Einstein, "Über Gravitationswellen," *Königlich Preußische Akademie der Wissenschaften zu Berlin. Sitzungsberichte*, January 31, 1918, 154–167.

6. B. P. Abbott and LIGO Scientific Collaboration, "Observation of Gravitational Waves from a Binary Black Hole Merger," *Physical Review Letters* 116, no. 6 (2016): 061102 (16 pages).

7. Alan F. Chalmers, *What Is This Thing Called Science?*, 4th ed. (Cambridge, MA: Hackett, 2013).

8. Jay L. Devore, *Probability and Statistics for Engineering and the Sciences*, 8th ed. (Stamford, CT: Cengage Learning, 2011).

9. Karl Popper, *The Logic of Scientific Discovery* (London: Hutchinson, 1968).

10. Chalmers, *What Is This Thing Called Science?*, 81–96; Lewis Wolpert, *The Unnatural Nature of Science: Why Science Does Not Make (Common) Sense* (Cambridge, MA: Harvard University Press, 1992).

11. Stephen Thornton, "Karl Popper," in *The Stanford Encyclopedia of Philosophy*, ed. Edward N. Zalta (Palo Alto: Stanford University Press, 2014) (online).

12. J. D. Bernal, *Science in History: Volume II, The Scientific and Industrial Revolutions*, 3rd ed. (Cambridge, MA: MIT Press, 1965).

13. Rom Harré, *Great Scientific Experiments: Twenty Experiments That Changed Our View of the World* (New York: Dover Publications, 2011).

14. Evelyn Fox Keller, *A Feeling for the Organism: The Life and Work of Barbara McClintock* (New York: W. H. Freeman, 1983).

15. David C. Cassidy, *Uncertainty: The Life and Science of Werner Heisenberg* (New York: W. H. Freeman, 1992).

16. Sindya N. Bhanoo, "Geology: Dinosaurs May Have Been Double-Teamed," *New York Times*, October 6, 2015, D4.

17. Niels Bohr, "On the Constitution of Atoms and Molecules, Part I," *Philosophical Magazine* 26, no. 151 (1913): 1–24; Niels Bohr, "On the Constitution of Atoms and Molecules, Part II, Systems Containing Only a Single Nucleus," *Philosophical Magazine* 26, no. 153 (1913): 476–502.

18. E. Schrödinger, "Quantisierung als Eigenwertproblem (Erste Mitteilung)," *Annalen der Physik*, 384, no. 4 (1926): 361–376; E. Schrödinger, "Quantisierung als Eigenwertproblem (Zweite Mitteilung)," *Annalen der Physik* 384, no. 6 (1926): 489–527.

19. Thomas S. Kuhn, *The Structure of Scientific Revolutions* (Chicago: University of Chicago Press, 1962).

20. Samuel A. Goudsmit, "Criticism, Acceptance Criteria, and Refereeing," *Physical Review Letters* 28, no. 6 (1972): 331–332.

21. Jeremy Berg, "Jiffy Pop," *Science* 353, no. 6299 (2016): 523.

22. David Wootton, *The Invention of Science: A New History of the Scientific Revolution* (New York: Harper, 2015); Bruce Alberts, Ralph J. Cicerone, Stephen E. Fienberg, Alexander Kamb, Marcia McNutt, Robert M. Nerem, Randy Schekman et al., "Self-Correction in Science at Work," *Science* 348, no. 6242 (2015): 1420–1422.

23. William Broad and Nicholas Wade, *Betrayers of the Truth: Fraud and Deceit in the Halls of Science* (New York: Simon and Schuster, 1982).

24. Ginger Pinholster, "Journals and Funders Confront Implicit Bias in Peer Review," *Science* 352, no. 6289 (2016): 1067–1068.

25. Paul Voosen, "Chemists Get Preprint Server of Their Own," *Science* 353, no. 6301 (2016): 740.

26. Kate Murphy, "Should All Research Be Free?," *New York Times*, March 13, 2016, SR6.

27. Tina Hesman Saey, "Repeat Performance: Too Many Studies, When Replicated, Fail to Pass Muster," *Science News* 187, no. 2 (January 24, 2015): 21–26.

28. B. A. Nosek, G. Alter, G. C. Banks, D. Borsboom, S. D. Bowman, S. J. Breckler, S. Buck et al., "Promoting an Open Research Culture," *Science* 348, no. 6242 (2015): 1422–1425.

29. John Bohannon, "Replication Effort Provokes Praise—and "Bullying" Charges," *Science* 344, no. 6186 (2014): 788–789.

30. Marsha McNutt, "Raising the Bar," *Science* 345, no. 6192 (2014): 9.

31. Marcia McNutt, "#Iamaresearchparasite," *Science* 351, no. 6277 (2016): 1005.

32. Mario Livio, *Brilliant Blunders from Darwin to Einstein: Colossal Mistakes by Great Scientists That Changed Our Understanding of Life and the Universe* (New York: Simon and Schuster, 2013).

33. Declan Butler and Jenny Hogan, "Modellers Seek Reason for Low Retraction Rates," *Nature* 447, no. 17 (May 2007): 236–237.

34. Felix Franks, *Polywater* (Cambridge, MA: MIT Press, 1981).

35. Nikolai N. Fedyakin, "Change in the Structure of Water During Condensation in Capillaries," *Colloid Journal USSR* 24 (1962): 425–430.

36. Boris V. Derjaguin, "Effect of Lyophile Surfaces on the Properties of Boundary Liquid Films," *Discussions of the Faraday Society* 42 (1966): 109–119.

37. D. L. Rousseau and S. P. S. Porto, "Polywater: Polymer or Artifact?," *Science* 167, no. 3926 (1970): 1715–1719.

38. P. Barnes, I. Cherry, J. L. Finney, and S. Peterson, "Polywater and Polypollutants," *Nature* 230, no. 5288 (1971): 31–33.

39. Jerry Suls and Rene Martin, "The Air We Breathe: A Critical Look at Practices and Alternatives in the Peer-Review Process," *Perspectives on Psychological Science* 4, no. 1 (2009): 40–50.

40. Ian Hacking, *The Social Construction of What?* (Cambridge, MA: Harvard University Press, 2000).

41. Ruth Bleier, "A Decade of Feminist Critiques in the Natural Sciences," *Signs* 14, no. 1 (1988): 186–195; Ruth Bleier, *Feminist Approaches to Science* (New York: Pergamon, 1988); Evelyn Fox Keller and Helen E. Longino, *Feminism and Science*, rev. ed., Oxford Readings in Feminism (New York: Oxford University Press, 1996); Helen E. Longino, *Science as Social Knowledge: Values and Objectivity in Scientific Inquiry* (Princeton: Princeton University Press, 1990).

42. John Bohannon, "Many Psychology Papers Fail Replication Test," *Science* 349, no. 6251 (2015): 910–911.

43. Stephen Curry, "Pespectives: It's Time for Positive Action on Negative Results," *Chemical and Engineering News* 94, no. 10 (2016): 34–35.

44. Kuhn, *The Structure of Scientific Revolutions*.

3 The Scientist and Truth: Dealing with Nature

Beauty is truth, truth beauty,—that is all
Ye know on earth, and all ye need to know.
—John Keats, "Ode on a Grecian Urn"[1]

Chapter 2 showed that the scientific quest for truth, for harmony between models and observations, is a formidable task. There are no infallible experiments, the models will be only imperfect glimpses of the ultimate truth, there will never be final and irrevocable answers. However, science has come close enough to truth for considerable success in predictive capability and practical applicability. After all, we can prevent polio and we can fly to the moon. How is it that science manages to work? What are the obstacles at each step, and how are they surmounted? What does this have to do with ethics?

In chapter 1, we set life, truth, the universe, knowledge, and justice as values in our ethical system. If we in any way compromise the pursuit of truth and knowledge, then we violate those values and undermine the progress of science. If we deliberately do *bad science*—for example, by faking data or by plagiarizing from the contributions of others—then we violate this value system. While we will discuss such bad science, we will focus on the daily decisions that test the integrity of those who are trying to do *good science*. We like to think that in science, we are fully objective in our work, that we seek the truth with open minds. But we are only human: we suffer from predilections, preconceptions, and prejudices that can keep us from the truth.

To organize our thinking, let us consider a scientific inquiry as having three steps: (1) the posing of a question or hypothesis about a phenomenon, (2) the making and analysis of observations on that phenomenon, and (3) the constructing of a model to organize the observations into an answer to the question or an assessment of the hypothesis, and thus to allow predictions of other phenomena. As discussed in chapter 2, these steps are convoluted, iterative, and interactive, but this outline provides a framework for thinking about the possibilities for error and bias in research.

What Is the Question or the Hypothesis?

It is considered a great compliment for one scientist to say of another that she or he "asks good questions." The very choice of the question should indicate that the issue is worth the time, expense, and effort required in its pursuit. The question necessarily reflects the context in which the scientist is working. The selection of one question can neglect or obscure other questions that may warrant attention. The manner of posing the question can constrain the approach to its solutions.

The context or *standpoint* of the scientist affects the questions that are posed.[2] The standpoint of the scientist will include:

1. The tools and technology available in a given time and place: James Watson and Francis Crick could not have determined the structure of DNA without the x-ray diffraction photographs of Rosalind Franklin.[3]
2. The previous progress in that realm of inquiry: Isaac Newton wrote, "If I have seen further ... it is by standing on ye shoulders of giants."[4]
3. The cultural milieu: Galileo tried to advance the idea of planets revolving around the sun at a time when the Catholic Church opposed that idea and imprisoned him for his pursuit of it.[5]
4. The source of funding for the research: a climate scientist whose work is funded by an oil company may find it hard to be objective.
5. The personal experience and imagination of the scientist: one scientist may ask a question that would not even occur to another scientist, which is a good reason for having a broad spectrum of people engaged in the scientific enterprise (see chapter 5).

The standpoints of scientists also limit their perspectives and can preclude asking important questions. The "unasked question" can mean that a variable that matters in understanding the phenomenon has been overlooked or ignored. An example of standpoint is a 2014 study that found an astonishing result: that the behavior of laboratory mice can depend on the gender of the person handling them![6] The mice respond to the odors of men, even odors left on their clothes. This result can have profound implications for a lot of work done over the years with laboratory mice, and raises questions about why the mice respond this way and whether there would be a different response to female handlers. No one had asked the question of whether the gender of the experimenter could affect the behavior of the mice.

When we pose a scientific question, we are already in danger of error, and we are ethically obligated to do our best to be aware of that possibility, to take it into account, to correct for it when possible, or to admit to it when we cannot correct for it.

What Are the Observations?

The late Professor Lothar Meyer of the University of Chicago used to say to his graduate students when the experimental results were puzzling: "Mother Nature is trying to tell us something."[7] We scientists have an ethical obligation to listen to Mother Nature, but there are subtle and perilous problems in hearing clearly what she saying to us.

How Do We Record and Store the Data?

The most famous and complex case of alleged scientific fraud of our time had at its core a failure to keep detailed and up-to-date laboratory notebooks and to record the primary data.[8] In 1990, molecular biologist David Baltimore was a Nobel laureate and president of Rockefeller University. In 1986, while he was at the Massachusetts Institute of Technology, he had published a paper with immunologist Thereza Imanishi-Kari and others on the genetics of antibodies. Imanishi-Kari's postdoctoral associate, Margot O'Toole, was later unable to reproduce the data in that paper, and accused Imanishi-Kari of having fabricated the data. Imanishi-Kari admitted to having kept disorganized records, but denied fabricating the data. Baltimore defended Imanishi-Kari. The issue was taken up by two scientists at the National Institutes of Health (NIH), Ned Feder and Walter Stewart, who believed O'Toole's allegations and took the issue to Representative John D. Dingell (D-Michigan), whose Energy and Commerce Committee had jurisdiction over NIH. In 1988, Dingell began an inquiry into the possible scientific fraud, subpoenaing documents and calling in the U.S. Secret Service to analyze the laboratory notebooks. The NIH Office of Scientific Integrity (OSI) also began an investigation. In 1991, under pressure from these investigations, Baltimore resigned as president of Rockefeller University and returned to the faculty of MIT. The OSI was replaced by the Office of Research Integrity (ORI) in the Department of Health and Human Services (HHS). In 1994, the ORI found Imanishi-Kari guilty of fraud and proposed barring her from government grants for ten years. She appealed the ruling and in 1996—ten years after the paper was published—she was finally cleared of all charges by the HHS appeals panel. Subsequent research has supported most of the findings of the 1986 paper. Had Imanishi-Kari made timely and complete records of her experiments and her data, a decade of humiliation and legal expense might have been avoided.

"Interpretations come and go, but data are forever."[9] Making available primary data is an ethical obligation of the scientist, an essential step in the replication of experiments that is key to scientific progress. The *raw* or *primary* data should be recorded before any corrections are made for background effects, calibration errors, and so forth. Those "corrections" could themselves turn out to be wrong and need to be redone.

There was a time when measurements were simply recorded in tables in paper laboratory notebooks. Today, most measurements are recorded on some kind of electronic medium. The development of electronic records raises important issues about how to preserve the information over many years, given that the nature of electronic media changes so rapidly. Many of us can remember 5.25-inch computer disks, which were superseded by 3.5-inch disks, which were superseded by zip disks, USB drives, and so on. Some of us can even remember paper tape, paper cards, and audio tape as storage media. If you had stored data on the earlier media, then you later had no easy way of retrieving that information.

Paper backup copies remain a reasonable long-term storage medium in many cases, but paper copies may not make sense for the really large data sets now being generated in some fields (e.g., climatology, x-ray crystallography, nuclear physics). Most journals now allow the submission of *supplementary information* to be a part of an accessible electronic record of an article. Moreover, the permanent storage of primary experimental data is becoming a requirement of federal funding agencies (see the websites for the National Institutes of Health and the National Science Foundation) and of scientific journals, in order that such data may be shared with other researchers.

Even if you have a computer record of the results of your experiments, you are well advised to keep a laboratory notebook. Your notebook can be a record of your motivations, your thinking, the nature of the samples used, the sequence of the operations, the failed experiments and negative results: a diary of the investigation that will prove invaluable later when you prepare your work for publication or for patent protection, or need to replicate the work as part of a new project. Appendix A gives guidelines for keeping a good laboratory notebook.[10] You will never regret extensive documentation.

Are There Systematic or Subjective Errors in the Data?

If you measure temperature with a thermometer, then you assume that the markings on the thermometer are correctly placed, meaning the thermometer has been properly calibrated. Let us assume that the thermometer was calibrated by putting it into ice water and into boiling water, and then dividing the degrees between the freezing point and the boiling point. If the ice water was dirty and that calibration point was wrong by one degree, then all measurements will have an error. This is an example of a *systematic error*, an error that affects all the measurements.

Systematic errors in instruments affect the *accuracy* of the data: the closeness of the measurements to standard scales of measurement. For temperature, this would be the agreement with the International Temperature Scale, as maintained at the U.S. National Institute of Standards and Technology. The *precision* of the data refers to the extent of the agreement when the measurement is repeated with the same instrument. It is possible to have high precision (say, 0.001 degree) but low accuracy (say, one degree), or vice-versa. Sometimes both high precision and high accuracy can be achieved. The precision may matter more than the accuracy if the goal is to take differences between values and absolute values are not important.

Sometimes systematic calibration errors can be corrected. You can check for calibration errors by measuring a known value. For example, you can test a thermometer by measuring the freezing point of distilled water. Then you can just add or subtract any error from all your thermometer readings. It is always wise to check instrument calibrations: it has been said that "[a]n experimenter of experience would as soon use calibrations carried out by others as he [or she] would use a stranger's toothbrush."[11]

You can get different kinds of systematic temperature errors if the experimenter tends to read the instrument too high or too low (a *subjective error*), if the probe is not in good contact with the system, or if the probe is so big that it perturbs the temperature. Vigilance is necessary in thinking about how measurements work.

Still other kinds of systematic errors can occur:

1. When the provenance of materials is taken for granted. For example, for many years biologists used cell lines that that were supposed to be human cells, but which turned out to be mouse cells.[12] For chemists and physicists, this means verifying the nature and purity of the materials under study.

2. When background signals must be subtracted in order to reveal the signals of interest. Then an incorrect background correction will lead to a systematic error. In the search for gravity waves, background signals from seismic activity, from cars, from cosmic dust, and so forth, have been serious impediments.[13]

3. When instrumental techniques give signals that are open to misinterpretation. A report of an image of a hydrogen bond from atomic force microscopy was later found to be an instrumental artifact.[14] A claim of a new supersolid phase of matter in helium was determined to be due to the much less interesting elastic properties of the solid.[15]

4. When computer programs used for data analysis have programming mistakes. In 2006, such a software error caused a crystallographer to retract five papers on protein crystal structures.[16] The primary data were still available, so he was able to reanalyze and republish. These problems, like instrument calibration problems, can be assessed by first using the program on a problem for which the answer is known, before using it on a new problem.

5. When there are *sampling errors*. The samples on which measurements are to be made must be randomly taken from the phenomenon of interest. If you are interested in studying the effects of changes in carbon dioxide level on the pH of Lake Michigan, then you must think carefully about the points in the lake from which you draw samples: not all samples can be taken near the highways or near the city of

Chicago. Only if you have truly randomized samples can you make use of the powers of statistical analysis.[17]

6.   When there is *experimenter expectancy*. When the experiment is a test of a theory and the experiment is designed in the context of the theory, then the experimenter may have an expectation that the data will fit the theory, and thus the data analysis is *theory laden*.[18] T. C. Chamberlin addressed this problem in 1890:

> The moment one has offered an original explanation for a phenomenon which seems satisfactory, that moment affection for his [or her] intellectual child springs into existence. ... So soon as this parental affection takes possession of the mind, there is a rapid passage to the adoption of the theory. There is an unconscious selection and magnifying of the phenomena that fall into harmony with the theory and support it, and an unconscious neglect of those that fail of coincidence. ... There springs up, also, an unconscious pressing of the theory to make it fit the facts, and a pressing of the facts to make them fit the theory. ... From an unduly favored child, it readily becomes master, and leads its author whithersoever it will.[19]

Systematic, subjective, and sampling errors are inherent to our scientific investigations. How do we ward against such errors? In 1830, Charles Babbage urged: "The first step in the use of every instrument [or computer program], is to find the limits within which its employer can measure the same object under the same circumstances. It is only from a knowledge of this, that he [or she] can have confidence in his measures of the same object under different circumstances, and after that, of different objects under different circumstances."[20] We must check measurement systems or computer programs for systematic errors, for stability over time, for repeatability by the same operator, and for reproducibility by a different operator. We can confirm the characterizations of all samples. We can try to convert any errors into random errors, which we know how to treat by the methods of statistical analysis. We can guard against imposing our expectations on the data. We then rely upon the community effort in science to check our work by replication and reproducibility, as discussed in chapter 2.

How Do We Analyze and Present the Data?

The analysis of the data begins before the data are collected. The experiment must be designed so that there are sufficient data of the right kind for the results to be statistically significant. The *signal* must be distinguishable from the *noise*. This implies an ethical obligation of the part of the scientist to understand statistics well enough to plan the data collection so that the signal is far above the noise, and then to calculate the appropriate significance levels and confidence tests that will convince other scientists of the results. Thus the first requirement of the data analysis and presentation is the initial design of a telling experiment: an experiment that will teach us something new.

Once the data are in hand, or even while the data are being collected, the scientist will plot a graph of the data to see an overall picture of the behavior.[21] A well-designed graph is a boon to thinking about the experiment and to presenting the results to the community of scientists.[22] Just as a good scientist learns statistics in order to plan and analyze the experiment, a good scientist studies the design of graphs: the selection of graph type (e.g., scatter, line, histogram), the use of shape and color to differentiate data, and the use of differences or ratios to amplify effects.

For example, figure 3.1 shows the deviation of the global mean surface temperature of the earth in the years 1850–2010 from the mean temperature for the years 1961–1990.[23] The mean temperature from 1961 to 1990 is subtracted in order to amplify the trends in the data. The year 1960 is at the zero of the ordinate, the years before 1960 have temperatures below that mean, and the years after 1960 have temperatures above that mean. Two data sets are shown in two different colors in the original paper. Bands give the 95 percent statistical confidence intervals for each data set. This well-designed graph makes the overall trend obvious to anyone: the mean temperature at the surface of the earth has increased over time. It is also obvious that there are large fluctuations in the data, fluctuations that are larger than the calculated confidence intervals. There is a lot of noise at the level of about ± 0.2 °C.

There are other considerations in presenting and analyzing data.

1. Extrapolation: We cannot make firm statements about behavior that is outside the range of the measurements. Some claim that examination of figure 3.1 for the most recent time period, 2000 to 2010, shows that the mean global temperature has remained constant for that period of time. This flattening has been used to assert that the temperature of the earth has stopped increasing, and to hypothesize the reason for that pause.[24] In fact, similar pauses can be seen in other time periods on the graph, for example, from 1960 to 1980. We may well ask whether the 2000 to 2010 pause is truly outside the general scatter in the data, and whether we can

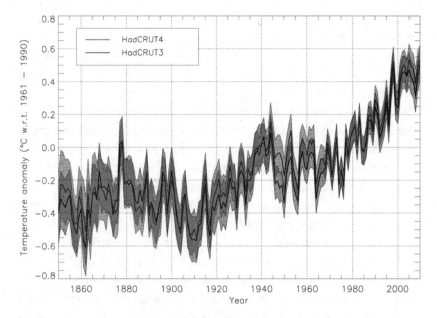

Figure 3.1

Deviation of the global mean surface temperature of the earth relative to the mean temperature in the years 1961–1990. The data include two data sets (HadCRUT3 and HadCRUT4). The bands indicate the 95 percent confidence intervals. *Source:* Colin P. Morice, John J. Kennedy, Nick A. Rayner, and Phil D. Jones, "Quantifying Uncertainties in Global and Regional Temperature Change Using an Ensemble of Observational Estimates: The HadCRUT4 Dataset," *Journal of Geophysical Research—Atmospheres* 117, no. D8 (2012): D08101 (22 pages), figure 7. With permission of the Journal of Geophysical Research. © Crown Copyright, Met Office.

extrapolate beyond the last year in the data set and assume that global warning has stopped. Such an extrapolation beyond the data, especially beyond data that are this noisy, is not justified.

2. Simplification: We lose information when we try to simplify the results of our measurements by presenting only an average or median value. An average or mean is the sum of the measurements divided by the number of measurements. A median is the measurement value that falls in the middle of the measured values. An average value can be distorted by one *outlier* that is somewhat larger or smaller than the other values. For example, assume that we measure the temperature of the air above the earth five times and get the values 55.0, 56.0, 54.0, 57.0, and 65.0 °F. The average is 57.4 °F. The median is 56.0 °F. The outlier at 65.0 °F has distorted the average value; the median gives a better measure of the temperature. If there are no outliers, then the average is more reliable. There are statistical methods for deciding whether an outlier should be discarded in calculating the average (see item 4, following).

An average or median value may be an essential first step, and may be interesting and useful, but that one number does not contain as much information as does the whole distribution of values. We need to know about outliers in the data set. We need to know about the extent of scatter in the data: any mean or average is useless without a measure of the deviation from the mean (the *variance*), such as the *standard deviation*. Once again: a good knowledge of statistics is essential for doing good science.

For figure 3.1, the investigators chose to plot a mean (average) temperature, taken over the whole earth. We might ask other questions of the data. Does the graph change if we look at data for the southern hemisphere as compared to the northern hemisphere? Are the temperatures over land masses different from those over the oceans, and how does that difference affect the mean? Has any difference between land temperatures and water temperatures changed since 1850? Looking beyond the simplest form of the data can reveal new insights.

3. Interpolation: We must take care in making statements about behavior that is between measurement points. For example, the density of water has a maximum at 4 °C.[25] If you have measured the density at 2 °C and at 6 °C, and then interpolated for the value at 4 °C, your value will be wrong and you will miss an interesting phenomenon.

4. Selection/omission: We must be careful when omitting data (outliers) that seem to be in disagreement with the rest of the data set. A scientist may know that, during an experiment, something was not right about the procedure—the power went out or a mistake was made—and can justify discarding data for that reason. However, when the data are collected and some data points look like they do not fit with majority of the other data, then the discarding of those outliers has to be carefully considered. There are statistical tests to justify omitting data, as will be discussed later in this chapter under the analysis of random error. It is not acceptable to omit data either to make the data look more precise (*trimming* the data) or to make the data fit an expected outcome (*cooking* the data).[26]

Since the advent of computer manipulation of images, it has become possible to alter data that are in the form of photographs and pictures. For example, an organic chemist may be tempted to remove peaks due to solvents and impurities from a spectrum of a newly synthesized compound before publishing the graph.[27] A biochemist may be tempted to "clean up" a photographic image from gel electrophoresis. Journal editors have made it clear that such changes in data images are not ethical: they are developing methods of testing images for manipulation, and they are sometimes requiring that original images or data be submitted. The 2016 website for *Science* instructs authors:

Science does not allow certain electronic enhancements or manipulations of micrographs, gels, or other digital images. Figures assembled from multiple photographs or images must indicate the separate parts with lines between them. Linear adjustment of contrast, brightness, or color must be applied to an entire image or plate equally. Nonlinear adjustments must be specified in the figure legend. Selective enhancement or alteration of one part of an image is not acceptable. In addition, *Science* may ask authors of papers returned for revision to provide additional documentation of their primary data.

There is another danger in discarding discordant data. Sometimes the outlier means that "Mother Nature is trying to tell us something." If we are too quick to dismiss the outlier, we can miss a chance to learn something or even to make a serendipitous discovery.[28] Indeed, many discoveries in science have resulted not from an exploration in the context of a hypothesis or a theory, but from an unexpected, anomalous observation that was further investigated. In 1938, German chemists Otto Hahn and Fritz Strassmann were puzzled when the bombardment of uranium with neutrons produced barium, an element of smaller atomic number than uranium.[29] They were not expecting such a result, even though it had been suggested from experiments by Enrico Fermi[30] and comments by Ida Noddack.[31] Then Lise Meitner and Otto Frisch figured out that the nucleus of uranium had split apart and formed barium.[32] Nuclear fission was discovered!

Are the Data Reproducible? Are Random Errors Properly Accounted For?

We discussed the issue of replication by other scientists in chapter 2, but we first must replicate our own work. When we make an observation in a scientific investigation, we are not finished until we are sure that we can repeat the observation, can verify the results, and can report on the level of uncertainty of the results. If we have accounted for known systematic errors (as discussed earlier), then the remaining error analysis will be that of the random error associated with the duplication of measurements. When we make the same measurement several times, the scatter in the data should be random and then can be analyzed by the methods of statistics. The statistical analysis will tell us whether some data points are outliers and can be discarded, but the reports should state that certain data points were omitted from the analysis. The statistical analysis will allow us to assign confidence intervals to the result in order to make meaningful comparisons with other measurements and with theoretical predictions.

The fact that experimental errors should be random was critical in the 2002 case of data fabrication by physicist Jan Hendrik Schön at Bell Laboratories of Lucent Technologies in New Jersey.[33] In 1999, Schön began publishing a series of papers, phenomenal in their number (almost one per week, totaling about ninety papers) and phenomenal in reporting one discovery after another of properties that could lead to making electronic devices out of organic materials. Schön had a distinguished educational background, he had highly regarded mentors and collaborators, he was at a premier research laboratory, and he published in the best refereed journals (including *Physical Review Letters*, *Nature*, and *Science*). His work was believed for three years, but all began to unravel because no one could replicate his experiments and because the data were too good to be true. He had kept no laboratory notebooks and he had destroyed all his samples, so it was hard to check his work. Finally, other scientists noticed that he had published the very same graph of data for three different phenomena in three different papers. The three data sets had the same random noise. Random noise is never reproducible: it is random! He had fabricated the data—three times. He was fired from his job, most of his papers were retracted, and his doctoral degree was rescinded. This case illustrates the role of random error and the success of self-correction in science, and it also raises interesting issues about collaborators and mentors for consideration in chapter 4: none of Schön's coauthors were held responsible for the fraud.

What Are the Models?

The questions posed and the data collected lead to models to explain the behavior of interest. The models can be qualitative or quantitative. The social and biological sciences use both qualitative and quantitative models. The physical sciences usually use quantitative or mathematical models.

The construction of models is subject to the same biases discussed previously for the posing of questions: biases in the minds of the scientists, and biases in the milieu in which they work. A bias in the mind of the scientist can lead in the right direction or in the wrong direction. For example, chemist Linus Pauling (figure 3.2), one of the finest scientists of the twentieth century, had succeeded in modeling proteins as helices in 1948.[34] His mindset (his standpoint) then led him to model deoxyribonucleic acid (DNA) as a helix, but he incorrectly postulated a triple helix that lacked acidic character. "Helices were in the air,"[35] so it was natural in 1953 for James Watson and Francis Crick to conclude from the beautiful x-ray studies of DNA by Rosalind Franklin and Raymond Gosling that DNA is a double helix. The "helices in the air" led Pauling to a wrong structure, but led Watson and Crick (with the additional x-ray evidence) to the right structure.

Often the comparison of a model to an experiment involves the fitting of a theoretically derived equation to the experimental data.[36] Usually this involves a *least squares analysis*: the free parameters in the equation are adjusted until the sum of the squares of the deviations of the data from the fitted line are at a minimum. This fit is more valid if the data points are weighted by dividing each value by its uncertainty, so that the better measurements count more than the worse measurements. In this case, a most useful graph is a graph of the *residuals*, the deviations of the data from the fitted line as a function of an independent variable. A residual plot amplifies the differences between the data and the model. For a good fit of the equation to the data, the residual plot should consist of random points, with no systematic deviations. If there are systematic deviations, then either there is a problem with the data or there is a problem with the model.

Research Conduct and Misconduct

In some of the cases discussed previously, scientists made mistakes that were honest ones: the question of supersolid helium, the misinterpretation of an image of a hydrogen bond, the erroneous identification of cell lines, Pauling's wrong structure for DNA. Albert Einstein himself was wrong

Figure 3.2
Chemist Linus Pauling (1901–1994) in 1954, with a model of the protein α-helix. Courtesy of the Special Collections and Archives Research Center at Oregon State University.

about a number of things: the existence of black holes, the expansion of the universe, the value of quantum mechanics.[37] In all these cases, scientists were working with integrity, but they made human errors. Scientists must have the liberty to be courageous in their work, to be able to report unexpected new results, to err without being punished. John Henry Newman said, "Nothing would ever be done at all if a [person] waited until [he/she] could do it so well that no one could find fault with it."[38] Chapter 2 explained the mechanisms for correcting such honest errors. Scientists who make honest errors are generally forgiven by the community, but a reputation for carelessness is to be avoided.

However, there are those in science who deliberately behave in dishonest and deceptive ways, as did Jan Schön (discussed earlier) and as have others.[39] An example of an experimentalist deliberately fabricating data to fit a theory is that of Emil Rupp, a German experimentalist working in the 1920s who created data to support Albert Einstein's theories of the nature of light, data that were then taken seriously by Einstein and others.[40] Rupp later recanted his work, claiming that mental illness led him to do it. In China, fraudulent behavior has included the selling of scientific manuscripts.[41]

In the United States, a report in 2005 found that among 3,247 scientists funded by NIH:

27.5 percent admitted "inadequate record keeping related to research projects";

15.3 percent dropped "observations or data points based on a gut feeling";

13.5 percent admitted "using inadequate or inappropriate research designs";

10.8 percent withheld "details of methodology or results in papers or proposals";

15.5 percent changed "the design, methodology, or results of a study in response to pressure from a funding source";

12.5 percent overlooked "others' use of flawed data or questionable interpretation of data";

6 percent failed "to present data that contradict one's own previous research."[42]

Only 0.3 percent admitted to falsifying data, but a striking 33 percent admitted to at least one of the behaviors listed. Another study in 2008 found similar rates of serious scientific misconduct.[43] The conclusion of the 2005 study was that "mundane 'regular' misbehaviors present greater threats to the scientific enterprise than those caused by high profile misconduct cases such as fraud."[44]

Why would scientists behave dishonestly? The reasons are like those for other human misbehaviors: laziness, ambition, envy, desperation.[45] Even if the fraudulent results are eventually ferreted out by the scientific community, that process is very expensive in time, energy, and money. In the 1990s, efforts to educate science students about ethical issues became more formal and more common. It can be hoped ethics training for students of science will, over time, raise awareness of ethical issues and reduce the incidence of misconduct.

Legal Definition of Research Misconduct

The federal requirements for training in research ethics came about as the result of several misconduct cases in the 1980s and 1990s. These cases also led to other new federal requirements and procedures for handling cases of research misconduct.

First, research misconduct is legally defined for the NIH in the Code of Federal Regulations (42 C. F. R. 93.103):

Research misconduct means fabrication, falsification, or plagiarism in proposing, performing, or reviewing research, or in reporting research results.

(a) Fabrication is making up data or results and recording or reporting them.
(b) Falsification is manipulating research materials, equipment, or processes, or changing or omitting data or results such that the research is not accurately represented in the research record.
(c) Plagiarism is the appropriation of another person's ideas, processes, results, or words without giving appropriate credit.
(d) Research misconduct does not include honest error or differences of opinion.

Fabrication, falsification, or plagiarism are often referred to as *FFP*. The *research record* in section (b) refers to laboratory records and all reports of the research (published papers, conference presentations, progress reports, etc.). Some issues (such as authorship controversies, human or animal subjects violations, and sexual harassment) are not included within FFP, but fall under other laws, regulations, and policies. Chapter 4 considers authorship and attribution issues; chapter 5 addresses sexual harassment, unlawful discrimination, human subjects, and animal subjects; chapter 6 includes intellectual property law.

Second, the NIH requirements for a finding of research misconduct are (42 C. F. R. 93.104):

(a) There be a significant departure from accepted practices of the relevant research community; and
(b) The misconduct be committed intentionally, knowingly, or recklessly; and
(c) The allegation be proven by a preponderance of the evidence.

Requirement (a) has been controversial because it is hard to establish what the "accepted practices" are, and because the deviation from accepted practices could signify creative, innovative new science that is to be encouraged. Requirement (b) may also be hard to prove, since it deals with motives that may not be discernable. Federal agencies other than the NIH have similar regulations, which can be found on their respective websites.

Procedures for Addressing Research Misconduct

As of 2000, the legal responsibility for policing research misconduct begins at the institution that has received federal funds.[46] Institutions that accept federal research funds are required to have procedures in place for handling accusations of research misconduct. A case is initiated when one person (the complainant or the *whistleblower*) alleges misconduct on the part of another person (the accused). The complainant reviews the institutional procedures and reports the accusation as required by the procedures.

At a university, often the first step is a conversation of the complainant with the vice-president for research (VPR), who will determine whether there is federal funding involved that will invoke the federal rules and whether the accusation seems credible. If the accusation needs to be pursued, then the VPR will begin an *inquiry* by assigning a committee or an individual to the task. If the inquiry indicates that that there is substance to the allegation, then there ensues a more formal *investigation*, at which point the matter must be reported to the funding agency: for NIH, to the U.S. Department of Health and Human Services Office of Research Integrity; for the National Science Foundation (NSF), to the Office of the Inspector General. The practice of keeping good records of primary data will serve the accusee well in the process. Whistleblowers, accusees, and investigators should carefully review the institutional procedures and the regulations of the relevant funding agency (on its website). If the accusee is found guilty by the investigation, then the accusee can ask for a review by an administrative law judge at HHS or by the Deputy Director at NSF, who then submits recommendations to the Assistant Secretary for Health at HHS or to the Director of NSF.

The institution can impose its own sanctions that may vary from a simple warning to termination of employment. The funding agency can add sanctions, including cancellation of existing grants, *debarment* (barring the accused from any federal funding for some period of time), retraction of published papers, and banishment from federal grant panels. Journals also can require the retraction of papers.[47]

Accusations of misconduct can also arise from outside the institution where the accusee is employed. For example, a reviewer of a paper or a proposal may find evidence of plagiarism. The accuser can then contact the journal editor or the agency program officer and the policies of the journal or of the agency will be followed in investigating the accusation.

Allegations of research misconduct are difficult for all involved. The whistleblower is protected by law from retaliation, but assuring that protection is difficult. The role of the whistleblower will be considered further in chapter 4. The accusee faces emotional distress and career damage, and can incur hundreds of thousands of dollars in legal fees. The institution itself is stressed by the time and energy dissipated in the inquiry. All are damaged by negative publicity in the scientific community. It is especially tragic if the accusee is found to be innocent, but has suffered greatly in the investigation and its aftermath.

Summary: The Scientist and Truth

The honest working scientist learns to be wary of the many pitfalls on the way to progress. Experiments have to be planned so that the signals exceed the noise. Raw data must be meticulously recorded. Observations are prone to systematic errors of several kinds. Extrapolation, interpolation, and oversimplification are to be avoided. Selection and omission of data must follow statistical tests. Random errors need to be properly treated by statistical methods. Theories are subject to bias and error.

Violations of good research practice are not uncommon, outright fraud is uncommon, but it is not possible to assess accurately the level of research misconduct. The community of scientists has mechanisms for correcting honest error and for addressing fraud, but the latter process is costly to science and to the public.

Physicist Richard Feynman urged scientists toward a "scientific integrity ... that corresponds to a kind of utter honesty—a kind of leaning over backwards ... to give *all* of the information to help others judge the value of your contribution."[48]

Guided Case Study on the Scientist and Truth: Self-Plagiarism

The publication by an author of the same text in more than one publication is called *self-plagiarism*.[49] In general, it is not acceptable to publish the same paper in more than one journal (duplicate publication), but exceptions will be examined in chapter 4.

Assume that it is not a whole paper that you want to publish again, but a section of a paper. Suppose that you have written a description of an experimental procedure and you have published that description in an article. Now you want to write a second article, in a different journal and with different coauthors, on a related experiment using the same procedure. Should you copy and paste the language from your first paper?

1. Name the values that are involved and consider the conflicts among them. Which values from chapter 1 are applicable? Is the text in the first paper also the intellectual property of the coauthors on that paper? What values enter because of the coauthors?

2. Would more information be helpful? Can you check the research integrity policies of the journals for relevant sections?

3. List as many solutions as possible. Can you just reference the earlier paper for experimental details (which forces readers to find the first paper in order to understand the second paper)? Can you copy the language, but put it within quotation marks and reference the first paper? Can you put aside the first paper and try to write out the experimental procedure in completely fresh language?

4. Do any of these possible solutions require still more information? Does the first journal own the copyright to your paper (see chapter 6 on intellectual property)? How does that affect your decision?

5. Think further about how you would implement these solutions. Do you need to talk to your coauthors? If you try to write a fresh version, does your mind keep coming back to the same language?

6. Check on the status of the problem. Has anything changed? Did your coauthors get back to you? Have you thought of another way to write that section?

7. Decide on a course of action. The safest course may be writing fresh new text, but that can be hard to do.

Discussion Questions on the Scientist and Truth

1. In 2015 a geneticist at the University of Chicago found an error in the analysis of a published paper and then publicized his new analysis on Twitter, the online news and social networking service based on 140-character messages, "tweets," rather than through the usual route of a reviewed correction in the original journal.[50] He argued that Twitter reached more people, more quickly. Who are the people who use

Twitter? What about appropriate review? What about permanent correction to the literature?

2. Discuss the advantages and disadvantages of electronic laboratory notebooks (ENLs). Is the rise of ENLs likely to increase or to decrease the keeping of effective records?

3. In medicine and psychology, there is a movement toward preregistering with a journal a research hypothesis and details about the statistical methods to be used to test that hypothesis.[51] The preregistration is expected to lead to more honest reporting of results. Does such preregistration make sense for physical scientists?

4. You are a synthetic organic chemist and you have just synthesized an important new compound. You ran the synthesis four times and you got four different final yields: 90 percent, 70 percent, 72 percent, 68 percent. What do you report as the yield for your procedure?

Case Studies on the Scientist and Truth

1. You have made a series of measurements on the phase diagrams of four liquid-liquid mixtures. The experimental procedures, the data analysis, and the comparisons to theory are the same in for all and the conclusions are the same in that they all support the prevailing theory. Should you report the work in one large paper or in four smaller papers? Of course, four papers may be more impressive to your department chair than just one paper. What are your considerations in thinking about this problem?

2. You are an analytical chemist in charge of the laboratory that measures the quality of the drinking water in your state. This is a new job for you, your first with supervisory responsibility, and you are still on probation. Your state is an economically depressed area and there are no funds available for improving the water purification system. You are reviewing the data taken over the past ten years and you are troubled by a sudden upward shift in lead contamination in the last two months. The shift coincides with the opening of a new battery plant that has provided quite a number of new jobs. What do you say to your subordinates? What do you say to your boss? How do you present the data? Lay out plans for addressing the problem, including both the case

in which your boss supports you and the case in which your boss does not support you.

3. You are a new assistant professor of materials chemistry, setting up your own laboratory. Two graduate students have already asked to work with you. You need to establish systems for cataloguing samples, recording experimental observations, and preserving calculations. What are your considerations? What instructions do you give to your students?

Inquiry Questions on the Scientist and Truth

1. In 1989, there was a report of the discovery of cold fusion.[52] What is cold fusion? How did the scientific community respond? What were the ethical issues? Was this a case of good science or fraudulent science? What is the current state of research on cold fusion?

2. Between 2004 and 2014, Professor Moses H. W. Chan and his coworkers first claimed the discovery of a new supersolid phase of helium-4, and then later proved that their original claim was mistaken.[53] This process is an example of the complexity of systematic errors in measurements, and also an example of good science at work, where the community of scientists raises questions respectfully and works with the original investigators to resolve the issue. Read more about this scientific investigation. Which came first, theory or experiment? What was the role of each in getting to the resolution? Is the issue fully resolved? Did this episode ultimately lead to new knowledge and new questions?

3. NIH requires that institutions receiving NIH grant funds have a procedure in place for dealing with research misconduct. What is that procedure for your institution? Do you think it is a good procedure? How might it be improved? Who should police misconduct in science, and why? What do you think should be the relative roles of Congress, NIH, NSF, professional societies, and university administrators?

4. Learn more about Galileo Galilei. What ethical issues did he encounter, and how did he handle them? What would you have done in his place?

5. From 1951 to 1969, physicist Allen V. Astin was the director of the National Bureau of Standards (NBS), now the National Institute for Standards and Technology (NIST) and an agency of the U.S. Department of Commerce.[54] During World War II, Astin had been a leader in the development of the proximity fuse, an important advance in

weapons technology. In 1952, NBS was asked to test a battery additive called AD-X2 which was supposed to make batteries last longer. The tests showed that the additive had no effect. Secretary of Commerce Sinclair Weeks disputed the NBS findings and forced Astin to resign as director. What happened after that? What were the ethical issues facing Astin?

Further Reading on the Scientist and Truth

Baykoucheva, S. *Managing Scientific Information and Research Data*. New York: Chandos Publishing (Elsevier), 2015.

Beveridge, W. I. B. *The Art of Scientific Investigation*. New York: Vintage Books, Random House, 1950.

Bleier, Ruth, ed. *Feminist Approaches to Science*. New York: Pergamon Press, 1986.

Braithwaite, Richard Bevan. *Scientific Explanation: A Study of the Function of Theory, Probability, and Law in Science*. New York: Harper and Row, 1953.

Charrow, Robert P. *Law in the Laboratory: A Guide to the Ethics of Federally Funded Science Research*. Chicago: University of Chicago Press, 2010.

Cleveland, W. S. *Elements of Graphing Data*. Murray Hill, NJ: AT&T Bell Laboratories, 1994.

Committee on Science, Engineering, and Public Policy. *On Being a Scientist: Responsible Conduct in Research*. 3rd ed. Washington, DC: National Academy of Sciences, 2009.

Derry, Gregory N. *What Science Is and How It Works*. Princeton, NJ: Princeton University Press, 1999.

Goodstein, David. *On Fact and Fraud: Cautionary Tales from the Front Lines of Science*. Princeton, NJ: Princeton University Press, 2010.

Harding, Sandra. *The Science Question in Feminism*. Ithaca, NY: Cornell University Press, 1986.

Hesman Saey, Tina. "12 Reasons Research Goes Wrong." *Science News* 187, no. 2 (Jan. 24, 2015): 24–25.

Ioannidis, John P. A. "Why Most Published Research Findings Are False," *PLoS Medicine* 2, no. 8 (2005): e124.

Judson, Horace Freedland. *The Great Betrayal: Fraud in Science*. Orlando, FL: Harcourt, 2004.

Karnare, Howard M. *Writing the Laboratory Notebook*. Washington, DC: American Chemical Society, 1985.

LaFollette, Marcel C. *Stealing into Print: Fraud, Plagiarism, and Misconduct in Scientific Publishing*. Berkeley: University of California Press, 1992.

Langmuir, Irving, and Robert N. Hall. "Pathological Science." *Physics Today* 42, no. 10 (1989): 36–48.

Lipson, Charles. *Doing Honest Work in College: How to Prepare Citations, Avoid Plagiarism, and Achieve Real Academic Success*. 2nd ed. Chicago: University of Chicago Press, 2008.

Oliver, Jack E. *The Incomplete Guide to the Art of Discovery*. New York: Columbia University Press, 1991.

Shrader-Frechette, Kristin. *Tainted: How Philosophy of Science Can Expose Bad Science*. New York: Oxford University Press, 2014.

Sigma Xi, The Scientific Research Society. *Honor in Science*. Research Triangle Park, NC: Sigma Xi, 2000.

Stennek, Nicholas H. *ORI Introduction to the Responsible Conduct of Research*. Washington, DC: U.S. Government Printing Office, 2004.

Taper, Mark, and Subhash R. Lele, eds. *The Nature of Scientific Evidence: Statistical, Philosophical, and Empirical Considerations*. Chicago: University of Chicago Press, 2004.

Taylor, John R. *An Introduction to Error Analysis: The Study of Uncertainties in Physical Measurements*. 2nd ed. Mill Valley, CA: University Science Books, 1996.

Tufte, Edward. *The Visual Display of Quantitative Information*. 2nd ed. Cheshire, CT: Graphics Press, 2001.

Waller, John. *Fabulous Science: Fact and Fiction in the History of Scientific Discovery*. New York: Oxford University Press, 2005.

Whitbeck, Caroline. *Ethics in Engineering Practice and Research*. 2nd ed. Cambridge, UK: Cambridge University Press, 2014.

Wilson, E. Bright. *An Introduction to Scientific Research*. New York: McGraw-Hill, 1952.

Notes

1. John Keats, "Ode on a Grecian Urn," *Annals of the Fine Arts* 4, no. 15 (1820): 638–639.

2. Sandra Harding, *The Science Question in Feminism* (Ithaca, NY: Cornell University Press, 1986).

3. James D. Watson, *The Double Helix: A Personal Account of the Discovery of the Structure of DNA* (New York: Atheneum Press, 1968); Brenda Maddox, *Rosalind Franklin: The Dark Lady of DNA* (London: HarperCollins, 2003).

4. Isaac Newton, "Letter to Robert Hooke, February 5, 1675, in the Digital Library of the Historical Society of Pennsylvania, record number 9792, Simon Gratz collection [0250A], box/case 12/11, folder 37.

5. Giorgio de Santillana, *The Crime of Galileo* (Chicago: University of Chicago Press, 1955).

6. David Grimm, "Male Scent May Compromise Biomedical Studies," *Science* 344, no. 6183 (2014): 461.

7. Sandra C. Greer, "Truth and Justice in Science," *The Hexagon of Alpha Chi Sigma* 87, no. 4 (1996): 67–69.

8. D. J. Kevles, *The Baltimore Case: A Trial of Politics, Science, and Character* (New York: W. W. Norton, 1998).

9. Marcia McNutt, "Data, Eternal," *Science* 347, no. 6217 (2015): 7.

10. Howard M. Karnare, *Writing the Laboratory Notebook* (Washington, DC: The American Chemical Society, 1985).

11. E. Bright Wilson, *An Introduction to Scientific Research* (New York: McGraw-Hill, 1952).

12. Rhitu Chatterjee, "Cases of Mistaken Identity," *Science* 315, no. 5814 (2007): 928–931.

13. Alexandra Witze, "Physics: Wave of the Future," *Nature News* 511 (2014): 278–281; Harry Collins, *Gravity's Shadow: The Search for Gravitational Waves* (Chicago: University of Chicago Press, 2004).

14. Jyllian Kemsley, "Hydrogen Bonds: Real or Surreal," *Chemical and Engineering News* 92, no. 51 (2014): 17.

15. Robert Halleck, "Is Solid Helium a Supersolid?," *Physics Today* 68, no. 5 (2015): 30–35.

16. Greg Miller, "A Scientist's Nightmare: Software Problem Leads to Five Retractions," *Science* 314, no. 5807 (2006): 1856–1857.

17. Charles Wheelan, *Naked Statistics: Stripping the Dread from the Data* (New York: W. W. Norton, 2013).

18. James Woodward and David Goodstein, "Conduct, Misconduct, and the Structure of Science," *American Scientist* 84, no. 5 (1996): 479–490; William Broad and Nicholas Wade, *Betrayers of the Truth: Fraud and Deceit in the Halls of Science* (New York: Simon and Schuster, 1982).

19. Thomas C. Chamberlin, "The Method of Multiple Working Hypotheses," *Science* 148, no. 3671 (1965): 754–759.

20. Charles Babbage, *Reflections on the Decline of Science in England, and on Some of Its Causes* (London: B. Fellowes, 1830).

21. W. S. Cleveland, *Elements of Graphing Data* (Murray Hill, NJ: AT&T Bell Laboratories, 1994).

22. P. Aarne Vesilind, "How to Lie with Engineering Graphics," *Chemical Engineering Education* 33, no. 4 (1999): 304–309.

23. Colin P. Morice, John J. Kennedy, Nick A. Rayner, and Phil D. Jones, "Quantifying Uncertainties in Global and Regional Temperature Change Using an Ensemble of Observational Estimates: The HadCRUT4 Dataset," *Journal of Geophysical Research—Atmospheres* 117, no. D8 (2012): D08101 (22 pages).

24. Isaac M. Held, "The Cause of the Pause," *Nature* 501, no. 7467 (2013): 318–319.

25. George S. Kell, "Precise Representation of Volume Properties of Water at One Atmosphere," *Journal of Chemical and Engineering Data* 12, no. 1 (1967): 66–69.

26. Babbage, *Reflections on the Decline of Science in England.*

27. Stephen K. Ritter, "Warning Shot on Data Integrity," *Chemical and Engineering News* 91, no. 25 (2013): 32.

28. Royston M. Roberts, *Serendipity: Accidental Discoveries in Science* (New York: John Wiley and Sons, Inc., 1989).

29. Ruth Lewin Sime, *Lise Meitner: A Life in Physics* (Berkeley: University of California Press, 1996); O. Hahn and F. Strassman, "Über den Nachweis und das Verhalten der bei der Bestrahlung des Urans mittels Neutronen entstehenden Erdalkalimetalle," *Naturwissenschaften* 27, no. 1 (1939): 11–15; J. Michael Pearson, "On the Belated Discovery of Fission," *Physics Today* 68, no. 6 (2015): 40–45.

30. Enrico Fermi, "Possible Production of Elements of Atomic Number Higher Than 92," *Nature* 133, no. 3372 (1934): 898–899.

31. Ida Noddack,"Über das Element 93," *Angewandte Chemie* 47, no. 37 (1934): 653–655.

32. Lise Meitner and O. R. Frisch, "Disintegration of Uranium by Neutrons: A New Type of Nuclear Reaction," *Nature* 143, no. 3615 (1939): 239–240.

33. Eugenie Samuel Reich, *Plastic Fantastic: How the Biggest Fraud in Physics Shook the Scientific World* (New York: Palgrave Macmillan, 2009).

34. Thomas Hager, *Force of Nature: The Life of Linus Pauling* (New York: Simon and Schuster, 1995).

35. Frances H. C. Crick, *What Mad Pursuit: A Personal View of Scientific Discovery* (New York: Basic Books, 1988).

36. John R. Taylor, *An Introduction to Error Analysis: The Study of Uncertainties in Physical Measurements*, 2nd ed. (Mill Valley, CA: University Science Books, 1996); Philip Bevington and D. Keith Robinson, *Data Reduction and Error Analysis for the Physical Sciences*, 3rd ed. (New York: McGraw-Hill, 2002).

37. Jeremy Bernstein, "Einstein: Right and Wrong," *New York Review of Books: NYR Daily*, January 9, 2016, http://www.nybooks.com/daily/2016/01/09/einstein-right-and-wrong/; Mario Livio, *Brilliant Blunders from Darwin to Einstein: Colossal Mistakes by Great Scientists that Changed Our Understanding of Life and the Universe* (New York: Simon and Schuster, 2013).

38. John Henry Newman, *Lectures on the Present Position of Catholics in England*, 1908 reprint ed. (London: Longmans, Green, and Co., 1851).

39. John D'Angelo, *Ethics in Science: Ethical Misconduct in Scientific Research* (Boca Raton, FL: CRC Press, 2012).

40. Jeroen van Dongen, "Emil Rupp, Albert Einstein and the Canal Ray Experiments on Wave-Particle Duality: Scientific Fraud and Theoretical Bias," *Historical Studies in the Physical and Biological Sciences* 37, no. Supplement (2007): 73–120; Jeroen van Dongen, "The Interpretation of the Einstein-Rupp Experiments and Their Influence on the History of Quantum Mechanics," *Historical Studies in the Physical and Biological Sciences* 37, no. Supplement (2007): 121–131; Jeroen van Dongen, *Einstein's Unification* (Cambridge, UK: Cambridge University Press, 2010).

41. Maria Hvistendahl, "China Pursues Fraudsters in Scientific Publishing," *Science* 350, no. 6264 (2015): 1015.

42. Brian C. Martinson, Melissa S. Anderson, and Raymond de Vries, "Scientists Behaving Badly," *Nature* 435, no. 7043 (2005): 737–738.

43. Sandra L. Titus, James A Wells, and Lawrence J. Rhoades, "Repairing Research Integrity," *Nature* 453, no. 7198 (2008): 980–982.

44. Martinson et al., "Scientists Behaving Badly."

45. Joeri K. Tijdink, *Publish and Perish: Research on Research and Researchers* (Amsterdam: Free University of Amsterdam, 2016).

46. Robert P. Charrow, *Law in the Laboratory: A Guide to the Ethics of Federally Funded Research* (Chicago: University of Chicago Press, 2010).

47. B. A. Nosek, G. Alter, G. C. Banks, D. Borsboom, S. D. Bowman, S. J. Breckler, S. Buck et al., "Promoting an Open Research Culture," *Science* 348, no. 6242 (2015): 1422–1425.

48. Richard P. Feynman as told to Ralph Leighton, *Surely You're Joking, Mr. Feynman: Adventures of a Curious Character*, ed. Edward Hutchins (New York: W. W. Norton, 1985).

49. Stephanie J. Bird, "Self-Plagiarism and Dual and Redundant Publications: What Is the Problem?: Commentary on 'Seven Ways to Plagiarize: Handling Real Allegations of Research Misconduct' (M. C. Loui)," *Science and Engineering Ethics* 8, no. 4 (2002): 543–544.

50. Yoav Gilad, "Open Review," *The University of Chicago Magazine*, July–August, 2015, 21.

51. Don A. Moore, "Preregister If You Want To," *American Psychologist* 71, no. 3 (2016): 238–239.

52. Gary Taubes, *Bad Science: The Short Life and Weird Times of Cold Fusion* (New York: Random House, 1993).

53. Halleck, "Is Solid Helium a Supersolid?"; E. Kim and M. H. W. Chan, "Probable Observation of a Supersolid Helium Phase," *Nature* 427, no. 6971 (2004): 225–227; Duk Y. Kim and M. H. W. Chan, "Upper Limit of Supersolidity in Solid Helium," *Physical Review B* 90, no. 6 (2014): 064503 (6 pages).

54. Walter Sullivan, "Allen V. Astin Is Dead at 79; Headed Bureau of Standards," *New York Times*, February 8, 1984, B14.

4 The Scientist and Justice: Dealing with Other Scientists

Science has its cathedrals, built by the efforts of a few architects and of many workers.
—G. N. Lewis and M. Randall, *Thermodynamics and the Free Energies of Chemical Substances*[1]

Science is done by the community of scientists, who build upon past work by others, who review one another's proposals and papers, who collaborate on projects, who teach and mentor one another, who leave a legacy as the foundation for the next generation. The success of science depends upon effective and ethical interactions within that community, and upon just behavior among its members.

Authors

A scientific investigation is not complete until the procedure and the results are made available to the community of scientists. Almost all physical and biological science is supported by funding entities (the federal government, private foundations, industry) and those entities expect and deserve formal reports for the funds they provide. Moreover, the progress of science requires that the work be made available for criticism and for extension. Scientists have an ethical obligation, to supporting agencies and to science, to publish the results of their research.

Scientists seek to publish in peer-reviewed journals, where their work will be carefully assessed by other experts in the field and where the published papers will be accessible to researchers around the globe. The publication of new, unreviewed results in newspapers or magazines that are not properly part of the scientific literature circumvents the vetting of the work by the scientific community and can lead to the propagation of mistaken science. An example of this kind of publicity in science is the 1989 case of the "discovery" of *cold fusion*—nuclear fusion at room temperature, as opposed to the nuclear fusion that happens at high temperatures in stars—by two scientists at the University of Utah.[2] Stanley Pons and Martin Fleischmann rushed to the newspapers to establish their priority in what turned out to be a case of erroneous, nonreplicable science, embarrassing themselves and their institutions and wasting the time and energy of those who had to sort out their claims.

A scientist who chooses to make public preliminary, unreviewed work—for example, on a personal webpage—should make clear that the research results have not been reviewed or are under review. Publication outside the peer-reviewed journals can also be justified if lives are in jeopardy, but it is hard to imagine that this would ever happen for most scientists.

How Much Work Constitutes a Paper?

French poet Paul Valéry wrote that "a [poem] is never finished, ... but [is] abandoned."[3] So it can seem with scientific research: a project could go on forever, but a scientist must eventually declare an end and publish the results. How does one decide when to publish?

There are two extremes in planning published papers. One is waiting too long, hoping for perfection, until the work loses interest and immediacy. The other is not waiting long enough and rushing into print with work that is incomplete or even mistaken.[4] Such judgments take thought, experience, and conversations with mentors and collaborators.

A good rule-of-thumb is that a scientific paper should make one good point, perhaps with one or two corollaries. Since published papers are the building blocks of a scientific career, there is an incentive to produce as many papers as possible. This can lead to the piecemeal publication of what is known as the *LPU* or *least publishable unit*, the breaking down of projects and results into as many papers as possible.[5] This practice is an inefficient use of the time and energy of the scientific community that must review and assess the work. It is better to seek to make each paper a complete "story." The National Science Foundation has tried to combat the impetus toward multiple publications by allowing only ten publications to be listed in a grant application.

There are situations when multiple publications of the same result are acceptable. Sometimes a researcher will publish a short report of an exciting new result in a *communications* or *letters* journal (such as *Physical Review Letters*). Such letters are very brief (perhaps three printed pages) and do not allow the presentation of full details. It is then incumbent upon the authors to provide a fully detailed report later in a "regular" journal (such as *Physical Review E*). However, that second, detailed paper often is never written. If the full details are never published, then replication by the scientific community is hindered. This lack of fully detailed papers was one of the problems in the 2002 fraud case of Jan Hendrik Schön, discussed in chapter 3.[6] Other acceptable forms of republication of the same material are reviews of the field (for example, in *Annual Reviews of Physical Chemistry*), articles aimed at the general public (for example, in *Scientific American* magazine), and articles that focus on the design of new instruments (e.g., *Review of Scientific Instruments*).

Who Are the Authors?

An author of a scientific paper or book must have made an *intellectual* contribution to the research. He or she could have had the idea and developed the research plan, or performed the experiments, or analyzed the data, or constructed a theoretical interpretation, or written the manuscript, or any combination of these. Most scientific papers have several authors, as will be discussed in the next section on collaborators. Usually the order in which authors are listed is in terms of the level of their contributions to the paper, with the person making the largest contribution listed as first author. However, the order of the listing of authors can depend on the particular scientific discipline. Even within the discipline of chemistry, the standard way of listing authors depends on the subdiscipline. Organic chemists tend to list the senior investigator (often a professor) *first*, while physical chemists list the senior investigator *last*. Sometimes the listing of authors is just alphabetical. The order of the listing of authors needs to be discussed early in the project in order to avoid hard feelings along the way.

Simply providing financial or administrative support is not enough to make a person an author. It is not acceptable to include *gift authors*— persons who really were not involved in the work. There is an amusing story that in 1948, the physicist George Gamow was preparing a paper with his student Ralph Alpher, and decided that it would be fun to add the name of colleague Hans Bethe, so that the authors would be "Alpher, Bethe, and Gamow," following the Greek letters alpha, beta, and gamma.[7] Bethe had not contributed to the paper, but allowed his name to be used; Alpher was not pleased.[8] While Gamow's sense of humor is to be appreciated, such loose use of authorship is not acceptable. It is common to thank others for the use of a piece of equipment, for helping with the figures, for editing the manuscript, for routine or paid technical support, but these colleagues are not to be listed as authors. All those who help with the project but who do not make an intellectual contribution can be listed in an acknowledgments section at the end of the paper. Scientists can be generous in acknowledging the help of others, while being careful about who gets the credit and the responsibility of authorship.

In general, those listed as authors are all held responsible for the content of the paper. Thus all authors must be included at each stage of the publication process, and no one should be listed as an author without her or his permission. Journals require that one author be designated the *submitting author*, to whom correspondence is addressed and who manages the submission and revision process for the paper. All authors should review the final paper before publication.

In contemporary science (see next section), often great strides can be made by using a team of scientists who have different areas of expertise. For example, a project on the thermodynamics of a protein might involve biochemists, experimental physical chemists, and theoretical physical chemists.[9] All the authors will not understand the details of all the parts of the project. Sometimes it may make sense to make a statement about a division of responsibilities at the end of the paper, but this is not yet common.

Collaborators

In the past, most science was done by one or two researchers, who performed every aspect of the work themselves.[10] Today, most science is done by teams of researchers: for example, a team of a professor with postdoctoral researchers, graduate students, and undergraduate students, or a team in which each member has a special expertise not shared by other team members.[11] Such *team science* is encouraged by the National Science Foundation and other funding agencies, which hold special competitions for support of interdisciplinary science. The reasons for the shift to team science are:

1. The need for special, expensive, complicated instruments that are not available everywhere and that are not easy to operate and interpret;
2. The need for several types of expertise to attack complex problems that span across disciplines;
3. The fact that scientific disciplines are themselves just human constructions: Mother Nature does not neatly divide her realm into chemistry and physics and biology.

Therefore scientists have moved from working alone at the laboratory bench to working in groups, with all the interpersonal and ethical issues that ensue. The issues of sharing authorship after the work is done were discussed earlier. There are also ethical issues in getting the work done: in making a fair division of labor and in monitoring the work to guard against error or misconduct. The team leader is responsible first for assigning the work among team members, and then for facilitating the communication and cross-checking within the team. Good coordination can make the roles and the assignment of credit clear from the beginning. Moreover, good oversight and communication can reduce the likelihood of error or fraud.

This raises the issue of who is to be held responsible in the case either of error or of fraud. In the case of error, the coauthors can participate in correcting the error (see chapter 2). In the case of fraud, the assignment of guilt can be complicated. When Jan Hendrik Schön (see chapter 3) was found guilty of falsifying data in some twenty-five papers on the electronic properties of organic materials, his twenty collaborators were not held responsible, except for their suffering the embarrassment of having the papers retracted from the journals in which they appeared.[12] Schön himself was fired from his job at Bell Labs and had his graduate degree withdrawn, but he was the only one who suffered such explicit punishment. The Schön case argues for the explicit listing of authorial responsibilities as discussed in this chapter, and led to the 2002 American Physical Society policy on coauthor responsibility (from APS website):

All collaborators share some degree of responsibility for any paper they coauthor. Some coauthors have responsibility for the entire paper. ... Coauthors who make specific, limited, contributions to a paper are responsible for them, but may have only limited responsibility for other results. While not all coauthors may be familiar with all aspects of the research ... , all collaborations should have in place an appropriate process for reviewing and ensuring the accuracy and validity of the reported results.

Predecessors

Each and every scientific paper is based upon the published papers of other scientists. Authors are ethically obliged to respect the contributions made by those previous workers by citing prior articles related to the new work. In the physical sciences, the citation usually is made by a numbered superscript that refers to a footnote or to a list of references at the end of the paper. (Each scientific journal has its own formatting requirements.)

The National Institutes of Health, via the Code of Federal Regulations (42 C. F. R. 93.103), defines *plagiarism* as "the appropriation of another person's ideas, processes, results, or words without giving appropriate credit." Thus plagiarism includes both using the exact words of another author without attribution, and also failing to give credit for the intellectual products of other scientists even if their exact words are not used (see chapter 6).

Sometimes it is useful to quote the exact words of another author. The simple and proper way to quote another person is to put those words within quotation marks and to provide a reference for the source. The paraphrasing of the words of another author can slide toward plagiarism, so it may be safer to quote directly than to try to paraphrase. Government funding agencies such as the National Science Foundation now use anti-plagiarism software to detect common language in grant applications, and journal editors are beginning to do the same for journal submissions.

Even if the exact words are not used, an intellectual contribution must be acknowledged. In order to acknowledge prior work, one must first be aware of that work. It is much easier to survey the literature in this day of search engines and electronic libraries than it was in earlier days of print journals. It remains hard to be aware of work in disciplines not one's own that is connected to the project at hand, but is perhaps written in a different vocabulary. For example, much of Albert Einstein's work on relativity— the constancy of the speed of light; the relativity of space, motion, and time; gravitational waves; $E = mc^2$—was anticipated in the 1902 work of the brilliant French mathematician Henri Poincaré, but Poincaré did not realize the implications for physics, and Einstein did not reference Poincaré's mathematical work.[13] A scientist's nightmare is that there is beautiful, relevant work somewhere in the literature, of which one is unaware.

It is necessary to cite the work of others that has preceded one's own work, even if the acknowledgment of that work makes one's own work seem not quite so original. French chemist Louis Pasteur made outstanding contributions to science, including the discovery of crystals that are mirror images of one another (optical isomers), the proof that life is not spontaneously generated, the advancement of the germ theory of diseases, and the development of vaccines for anthrax and rabies. However, Pasteur made a habit of not acknowledging, and even of disparaging, his predecessors, and he repeatedly failed to give credit to his own collaborators.[14] He did not acknowledge that August Laurent had already established the connection between crystalline form and optical activity.[15] In the famous 1881 trial to demonstrate the efficacy of the anthrax vaccine, Pasteur actually used a vaccine that was developed by Jean-Joseph Toussaint, but he did not reveal that this was the case. In his development of the vaccine for rabies, he usurped the preparation technique of his close collaborator Emile Roux. Pasteur was an ambitious scientist who put his own career ahead of the rights of his predecessors and collaborators.

There may be cases where the work you cite is so well known that you may not need to include the reference. For example, if you use "$E = mc^2$," do you really need to cite Einstein's paper from 1905,[16] or is the equation so much a part of our culture that it needs no reference? Or what about such constants as Avogadro's number, N_A? Is the value of N_A to be taken for granted, or should the scientists who determined that value be acknowledged?[17] It is always safer to include the reference.

Scientists regularly share their unpublished results at national and international meetings and conferences. The open exchange of information assumes that those present will respect the ongoing work of the presenters and will not use their ideas and results without permission.

It is termed *self-plagiarism*, and even fraud, if an author submits the same manuscript to more than one journal and seeks to get the same work published in several places. It is, however, acceptable to obtain a patent before submitting a scientific publication, since patents are not considered publications. The reuse of specific language and ideas should considered carefully in revising and resubmitting papers and grant proposals, especially if the coauthors have not remained the same, since the first paper or proposal was the intellectual property of all the original authors.[18] This problem is considered in the guided case study at the end of chapter 3.

Reviewers and Referees

As discussed in chapter 2, reviewers and referees are involved before and after a scientific project is done. Here again, *reviewer* will indicate review of both proposals and of manuscripts.

As noted, science is usually done with financial support from a government agency or a private foundation or company. Before the funds are allocated, several experts in the field review the proposed project, to determine whether there is a good scientific question, whether the principal investigator is qualified to do the work, and whether the plan for the work makes sense and is feasible. After the research is completed and a paper is submitted to a refereed journal, the journal editor selects two or three experts to decide whether the work is sound, and whether the manuscript meets standards for the level of communication, the depth of detail and analysis, and due credit to prior work. The granting officers and the journal editors depend on the reviewers to help make the decisions on whether to fund the research in the first place, and then whether to publish the results of the research.

The members of the scientific community who serve as reviewers do so as a service to the community, with no compensation. The motivation for serving is that the reviewer will also need to have his or her own papers and proposals reviewed, and thus the service will be reciprocated. Authors are allowed to recommend particular people to be reviewers and are allowed to ask that particular people not be reviewers. The use of single blind and double blind review was discussed in chapter 2. Usually the journal reviewers work independently, but recently some journals are allowing *cross-review* in which the reviews are exchanged among reviewers for a more complete analysis before the reviews are sent to the authors.[19] Similarly, grant proposals are often studied by panels of reviewers who can discuss and compare the proposals. The final decision on whether to publish the paper or to fund the proposal is made by the editor or the program officer, on the basis of the reviews.

The first obligation of a reviewer is to respond in a timely way. It is not acceptable to let the task wait for weeks. The next obligation is to give serious attention to the paper or proposal because there is a lot at stake. First, a proposal represents, in itself, an investment of time and effort by the proposers, and second, the proposal requests an allocation of resources that may (or may not) be better invested elsewhere. A scientific paper represents months and even years of work by the authors, and can be a step in the progress of science. The reviewers have the opportunity to keep the literature from being polluted by erroneous or fraudulent work. A reviewer can ask that the authors provide more details, or more data, or more analysis.[20] A reviewer can check that the reference list does not omit important prior papers. In any case, the responses of the reviewer should be courteous and respectful, offering constructive criticism and never destructive comments. Anonymity is not a license for rudeness or meanness.

The reviewer is expected to keep the paper or proposal confidential. It is not acceptable to give copies of the paper to others or to make use of the paper in one's own work without the permission of the author. If issues arise about appropriate use of the manuscript, the reviewer should consult the grant officer or journal editor about how to proceed. The reviewer should not contact the authors directly without the approval of the officer/ editor, because this violates the confidentiality of the review process and interferes with the oversight role of the editor/officer.

It can happen that a reviewer has a conflict of interest in a particular case. A *conflict of interest* (see chapter 6) means that the reviewer cannot make an unbiased assessment because of a relationship to the authors or to the material. On the one hand, the authors may be friends or relatives, or former students, or collaborators of the reviewer. On the other hand, the authors may be rivals or competitors of the reviewer. The substance of the paper or proposal may overlap with work that the reviewer is pursuing. Some conflicts are inevitable, since the reviewers and the authors necessarily work in the same field. The reviewer faces the ethical issue of whether there is even an appearance of improper bias, or whether she or he can review the paper or proposal with fairness and objectivity. The reviewer can always consult with the editor of the journal or the program officer handling the proposal, so that they can decide together whether the reviewer should proceed or be replaced.

Grantees

When a scientist seeks funding for a research proposal from funding agencies and foundations, there are ethical issues at every stage of the process. A researcher may submit duplicate or very similar proposals to different funding entities, but must inform all agencies of the initial duplication and then of any funding that is awarded. It is not acceptable to accept financial support from more than one source for the same research.

Once a grant is received, it must be spent for the purpose for which it was requested. Of course, there is a good deal of flexibility in any work plan for a research project: you do not know ahead of time how it will go and you may need to change the plan—or it would not be research! However, any significant change in the purpose or scope of the work or in the allocation of funds must be approved by the program officer. For a contract (as opposed to a grant), there is less much flexibility because specific products are expected at the end of the contract period.

The interim reports and the final report must be submitted on time, or the funding will be in jeopardy. These reports will include descriptions of progress on the project, and those reports should be scrupulously honest, resisting any temptation to exaggerate the level or significance of the results.

Journal Editors and Grant Officers

Editors and grant officers have their own ethical challenges. They are themselves members of the scientific community, with inevitable conflicts of interest in handling the papers and proposals of their colleagues.

Editors are responsible for managing and developing their journals. Journals are judged by the journal impact factors (JIF, see chapter 2). For example, a prestigious journal such as *Science* will have a JIF of about 30, whereas a more focused journal may have a JIF close to 1.0. Editors may be tempted to enhance the impact factors of their journals by encouraging authors to cite their own journal more frequently, or by publishing more review articles to garner more citations.[21] Such efforts can be detrimental to the goal of publishing good science and promoting fair citation practices.

Whistleblowers

What are the ethical obligations of a scientist who becomes aware of a case of error or a case of misconduct on the part of other scientists? When should the scientist become involved in correcting the error or reporting the misconduct? The scientist must first decide whether the problem is an error or is misconduct, and then decide what to do about it. It is essential to be absolutely sure about a determination that misconduct has happened, because a mistaken judgment can bring dire consequences for all, including the whistleblower. The seven-step ethical decision-making process for making ethical decisions given in chapter 1 will serve well in thinking through these decisions.

If the case is one of human error rather than misconduct, then the courteous and respectful action is to contact the other scientists personally and thus allow them the chance to correct the error themselves. If the other scientists disagree about the reported honest error, then this honest scientific disagreement will be resolved by further discussion and research, as discussed in chapter 2.

If the case appears to be one of unethical or illegal conduct, then the scientist—now the *whistleblower*—must think carefully about how to proceed. If there is misconduct, then failure to report is collusion in the misconduct.[22] It may make sense to discuss the matter confidentially with another person before going to the official who handles such cases in the organization.[23] That person may be a mentor to the whistleblower, or may be in a position to know more about the matter at hand, and may be able to support or refute the determination of misconduct. The whistleblower must gather as much information and documentation as possible, and then proceed—with professional respect toward all—to those who have the authority to review the case. A procedure for reporting research misconduct, as discussed in chapter 3, should be in place in each organization. If for some reason the procedure is not clear, then the whistleblower will need to start with the immediate supervisor and, if necessary, move upward through the organizational chain of command.

It is wise to keep a charge of misconduct confidential at every step, and to keep a dated diary of each conversation in the process. Avoid using electronic communications (email or text) for any of these interactions, because such communications can be subpoenaed in the event of legal proceedings.

A whistleblower can be in danger of career damage, even if the charges are correct.[24] Whistleblowers are sometimes seen as disloyal to colleagues and to the organization. There can be pressure to drop the charges because of the expense and bad publicity. There can accusations of bad faith on the part of the whistleblower, ostracism by colleagues, loss of support from superiors. The U.S. Whistleblower Protection Act of 1989 and later additions to this act provide protection against retaliation to persons who report illegal or unethical behavior.[25]

Whistleblowers play an important role in maintaining the integrity of science. They may be the only people aware of a wrongdoing, the only ones in a position to stop the behavior. They deserve the respect and protection of the scientific community.

Role Models, Advisors, Supervisors, and Mentors

A *role model* is someone who sets an example for others, but who is not necessarily personally connected to those others. For example, Marie Sklodowska Curie can be a role model for junior (and senior) scientists, exemplifying love of science and dedication to research, even though she is no longer living. An academic *advisor*—a research advisor or thesis advisor—directs an undergraduate or graduate student or a postdoctoral associate toward mastery of the skills needed for success as a scientist and toward completion of the academic degrees needed for a career in science. A *supervisor* is a person with formal administrative authority over another person or group—for example, an industrial research group director supervises scientists in that group, or an academic department chair supervises the faculty members in that department.

A *mentor* is someone who rises to a higher level than either advisor or supervisor by assuming the obligation to help another person—the *mentee*—to find her or his way, and by making the welfare of the mentee a high priority. Mentors are important in personal lives (parents, friends, counselors) as well as in professional lives, but the focus here is on mentors in professional lives. The relationship can be formal: sometimes administrators assign mentors to junior colleagues. The relationship can be informal: a female graduate student may seek mentoring from a female faculty member who is not her research advisor. The key to the role of mentor is an abiding interest in the welfare of the mentee, "coaching, ... nurturing, ... sponsoring, ... [and] tutoring."[26] Mentors willingly assume the ethical responsibility for advancing the professional lives of their mentees.

There can be overlap in these roles: an advisor can also be a mentor or a role model, for example, but this does not necessarily happen. A mentor will necessarily be a role model, but a role model need not be a mentor. This discussion centers on the general category of mentor, assuming some overlap with the roles of advisor and supervisor.

Scientists find themselves both mentoring and being mentored. Mentors may include direct supervisors, but will often include other people who are willing and able to serve as mentors. Indeed, mentoring can be among scientists at the same level of professional development (*horizontal* mentoring[27]) who meet to help one another with problems. Mentoring is seen as important to increasing the participation of women and minorities in science and engineering (see chapter 5). For example, the Association for Women in Science (AWIS) has developed webinars and workshops to train mentors (see AWIS website).[28]

In an industrial or government setting, the manager at each level may be a mentor for her or his supervisees. William Shockley was the supervisor of John Bardeen and Walter Brittain at Bell Laboratories when Bardeen and Brittain invented the transistor in 1947.[29] Schockley had nothing to do with the invention, but he wanted a part of the glory. He took over control of the project, and then developed the junction transistor, using ideas from Bardeen and Brittain. The Nobel Prize in Physics was awarded to all three men in 1956. A good mentor will advance the welfare of the mentee and will not take unfair advantage.

In an academic setting, the research advisor frequently serves as a mentor to members of his or her research group. An example of brilliantly successful mentoring centers on the British crystallographers in the first half of the twentieth century.[30] William Henry Bragg (1862–1942) and his son William Lawrence Bragg (1890–1971) had developed the technique of using x-rays to determine the structures of crystals, for which they won the Nobel Prize in Physics in 1915. William H. Bragg mentored Kathleen Lonsdale (1903–1971), and invited her back to science after she spent time raising three children; she was later the first woman to be elected a Fellow of the Royal Society and to serve as president of the International Union of Crystallography. William H. Bragg also mentored John Desmond Bernal (1901–1971), who himself mentored (among others) Dorothy Crowfoot Hodgkin (1910–1994), Aaron Klug (1926–), Rosalind Franklin (1920–1958), and Max Perutz (1914–2002). W. Lawrence Bragg was a mentor to Francis Crick (1916–2004) and to James D. Watson (1928–).

Almost all of these crystallographers won Nobel Prizes. Rosalind Franklin, who made the x-ray photographs that were key to unraveling the structure of DNA, died before the Nobel Prize in Physiology or Medicine was awarded to James Watson, Francis Crick, and Maurice Wilkins in 1962.[31] They were a diverse group of people: Klug, Franklin, and Perutz were Jewish; Lonsdale, Hodgkin, and Franklin were female; Hodgkin was coping with rheumatoid arthritis and raising three children. It was a phenomenal mosaic of mutually supportive relationships that resulted in much great science. (The one big exception was the behavior of Watson and Crick, who used Franklin's x-ray photographs of DNA without her permission.[32]) The Braggs set an example of mentoring, and their example was bequeathed to their scientific descendants.

Guidelines for Mentors

Mentoring includes career needs such as "coaching, network building, fostering opportunities, and facilitating exposure," and also includes mentoring for psychological and social needs such as "emotional support,

Figure 4.1
Research group of W. H. Bragg (1862–1942) (in center) in 1930. Photo by Serge Lachinov. This work is in the public domain.

counseling, role-modeling, and friendship."[33] Successful mentors need to be experienced, knowledgeable, empathetic, insightful. They need to have high standards for personal and professional conduct, to appreciate the importance of diversity in science (chapter 5), and to be willing to give time and attention to mentees.[34]

Mentors generally are persons with more career power than the mentees, so that a part of the mentoring process is the sharing of that power to benefit and develop the mentee. Mentoring in science and engineering often will mean that the mentor and the mentee are from different demographic groups. For example, a white male professor may mentor an Asian female student. Such mentor-mentee relationships do succeed when the mentor is knowledgeable about and sensitive to the differences of experience and culture, but the mentee may still need other support for psychological and social mentoring.

Career mentoring means that the mentor will help the mentee to reach her or his career potential. If you are a professor-mentor, that means advising your student mentees along paths that make the best use of their talents, even if those path are at odds with your own needs or views. It means guiding them in their work so that they achieve an academic degree in a reasonable period of time, tracking their progress, weekly if not daily, being available and attentive. It means providing for your mentees to attend conferences, promoting them for invited talks, and recommending them for job opportunities.

Psychological/social mentoring will include helping mentees to understand the social structure and mores of the scientific community, the norms and expected behavior. Mentees from underrepresented groups (see chapter 5) may need special social mentoring to survive and prosper. Mentors can also help to mitigate the stresses of a scientific career, from graduate student struggles, to tenure decisions, to grant denials, to work/family conflicts.

Laboratories present dangers from chemicals, radioactive samples, biological materials, lasers, mechanical devices, and more. Supervisors and mentors of scientists must insist that safety procedures and regulations be followed. For example, the 2009 death of a research assistant in organic chemistry at the University of California at Los Angeles was due to improper use of a flammable chemical.[35] The assistant died of injuries from the fire, the professor's career was derailed, and the university spent $4.5 million in legal fees. The U.S. Department of Labor Occupational Safety and Health Administration (OSHA) sets requirements and guidelines for laboratory safety. Research supervisors must understand OSHA requirements, institute safety training, provide protective equipment, establish standard operating procedures, and promote awareness of hazards. Attention to safety is respect for the value of life, as declared in chapter 1.

Lastly, role models, advisors, supervisors and mentors share the leadership responsibility for educating in all the aspects of professional ethical behavior that are considered in this book—data management, interpersonal relationships, grant management, societal implications of research, and all the roles discussed in this chapter. It is the responsibility of institutions to provide training for employees so that they are prepared to serve in these roles.

Guidelines for Mentees

Mentees need to take the responsibility for getting the mentoring that they need. Each will generally need more than one mentor at any point in time, and will need to shift mentors as circumstances change. The need for mentoring will persist over an entire career, but the nature of the mentoring will change. Even retirees need the advice of others who have made that transition.

Mentees should expect from mentors the kinds of support already discussed. Indeed, mentees may need a mentor to guide them in finding other mentors: a meta-mentor! For students, a department chair can be such a meta-mentor; in industrial and government settings, the human resources department may be of help.

Summary: The Scientist and Justice

An undergraduate chemistry student was overheard to say that she did not want to be a scientist because scientists stay alone in their laboratories and never interact with other people. This is not the case. Science is a human endeavor that depends on human interactions. Scientists build upon the work of their predecessors, work with one another as collaborators and reviewers, and teach mentees for the next generation. In all these relationships, we want to treat one another with civility, kindness, and fairness, helping one another to do the very best science, and seeking to give credit and support to each other.

Guided Case Study on the Scientist and Justice: Whistleblowing

Your colleague Jeff has just attended a meeting of the American Chemical Society in Orlando, Florida, with his wife and two children along. You know that he charged his NIH grant for all his expenses. Jeff brags to you that he did not attend the meeting at all, but instead he spent the entire time at Disney World with his family. What should you do?

1. Name the values that are involved and consider the conflicts among them. Revisit the list of values in chapter 1. Are there also legal issues?

2. Would more information be helpful? Are you sure that he used NIH funds for all expenses? Are you sure that he attended no sessions? Was he scheduled to give a talk? Can you find out more about illegal uses of grant funds?

3. List as many solutions as possible. Should you say something to Jeff? To your department chair? To a more senior colleague or mentor? To the vice-president for research?

4. Do any of these possible solutions require still more information? Has your institution published standard procedures for reporting misconduct?

5. Think further about how you would implement these solutions. What will your other colleagues think about Jeff? About you? What if Jeff denies the accusation?

6. Check on the status of the problem. Has anything changed? Has Jeff filed for expense reimbursement? Can your department chair review that claim?

7. Decide on a course of action. Review the earlier section about whistleblowing, especially the parts about maintaining confidentiality, keeping a diary, and not using email.

Discussion Questions on the Scientist and Justice

1. Is it unethical to cite an article or book that you have not read?

2. Often a scientist will write a *review article* that gives an overview, a summary, and a synthesis of the current status of work in a particular field. The review will include references to all the primary work in the field. Later papers in the field may refer to the review article and not refer to the primary literature. Is this disrespectful to those who did the primary work? Is it unethical?

3. Robert Merton has proposed a *Matthew effect* in science.[36] The Holy Bible (King James version), in Matthew 13:12, says: "For whosoever hath, to him shall be given, and he shall have more abundance: but whosoever hath not, from him shall be taken away even that he hath." Merton proposes that, on the one hand, scientists who are famous get all the credit for papers on which their names appear, even if they made minor contributions. On the other hand, significant contributions by unknown scientists are often ignored. What does this say about the review system? About justice in science? What can be done about it?

4. It has been said that "part of the intellectual responsibility of a scientist is to provide the best possible case for important ideas, leaving it to others to publicize their defects and limitations."[37] Do you agree? Do you think that this is what happens?

5. It is common in the social sciences to explain the ordering of the names of the authors of an article in a footnote. For example, the note may say, "The names of the authors are given in alphabetical order," or "The names of the authors are in the order of the level of contribution to the project, the first author being the major author." Such explanations are not common in the physical sciences. Should physical scientists also make the contributions of authors explicit?

Case Studies on the Scientist and Justice

1. Juan is a graduate student in materials chemistry. In order to complete his dissertation research, his advisor, Professor Lee, makes arrangements for him to use a differential scanning calorimeter in the laboratory of another faculty member, Professor Random. Random agrees that several samples may be run on his instrument, in exchange for which Lee graciously supplies a new box of sample cells for the calorimeter supply drawer. Juan does his measurements over the weekend, so as to minimize his interference with the ongoing work in Random's laboratory. When the work is written up for publication, Random sees the preprint and insists that his name be added as an author. Should Random be an author on the paper?

2. Graduate student Debra has collected data on neutron scattering from a magnetic solid. The sample is hard to prepare, so such studies have never been done before. She writes her PhD dissertation and also

publishes two papers from her data analysis. After she has graduated and has changed fields of research, a new graduate student in the same laboratory, Sophia, reanalyzes Debra's data and obtains an interesting new result.

    a. If the original data were *not* published as supplements to Debra's papers, but were still available in the laboratory files, should Debra be a coauthor on Sophia's paper?

    b. If the original data *were* published as supplements to Debra's papers, should Debra be a coauthor on Sophia's paper?

    c. If Debra also took some data on a second compound and did not analyze or publish those data, but Sophia has analyzed these unpublished data and prepared a second paper on this compound, should Debra be a coauthor on this paper?

3. Janice is a materials physicist in an industrial laboratory. She has an idea that a particular organic compound will have semiconducting properties. She asks George, an organic chemist, to prepare the compound for her and he agrees to do so.

    a. If George does not understand the physics of the project, should George be a coauthor on the resulting paper?

    b. What if instead George had realized the promise of a new compound and asked Janice to make the electronic measurements? Should her name be on that paper?

4. Professor Bruger agreed to allow a high school student, James, to spend the summer working in his laboratory. James worked along with a postdoctoral researcher, learning to operate the instrument and to collect data. However, James did not have the background to understand the project fully and he could not help with the analysis of the data. Should James be a coauthor on the paper? How does this case compare to problem 3?

5. Georgina is a young and unknown scientist, making her career at a government laboratory. Albert is a senior and famous scientist who has the adjacent office and who works on similar scientific issues. After hearing that Georgina has an exciting new result, Albert comes to her office and offers to be a coauthor on her paper. Albert says that, with his name on the paper, it will be more likely to be accepted by a good journal and will read by more people. Should Albert be a coauthor?

6.   Alex, an established researcher in industry, is looking through a new
     issue of a journal and he sees an article that is closely related to earlier
     work of his own, but that does not cite his papers. What should he do?
     Contact the authors? Contact the journal? Write a Comment on the
     article?

7.   Donna attends a summer Gordon Conference on the Physics and
     Chemistry of Liquids, where she hears a lecture on unpublished work
     on phase transitions under pressure. She realizes that the information
     in the lecture will explain some data over which she has been puzzled.
     Can she use that information to complete her own paper?

8.   Edward has a grant from the National Institutes of Health to work
     on electron transport in DNA. In the midst of this work, he realizes
     that he can apply his techniques to electron transport in neurons.
     Can he "bootleg" the neuron work on his NIH grant? What are his
     options?

9.   Ellen is reviewing an article for the journal *Science*. She notices that the
     authors have not referenced a paper of her own that is relevant to their
     work. Is it a conflict of interest for her to advise the editor that this
     reference be included? What if her paper is only marginally relevant to
     the new paper?

10. The rule in Professor Sundhurst's laboratory is that no one must ever
     work alone in the laboratory. You are working in the lab late one night
     and you find that you must go home to help your spouse with a sick
     baby. One other student is in the lab and is in the midst of a complex
     preparation that will take at least another two hours. You cannot leave
     her alone. What do you do?

11. Puijun reads a new paper in *Physical Review Letters* and realizes immedi-
     ately that the authors have overlooked some information that negates
     the point of the paper. What should she do next? What if the authors
     of the paper do not agree with her point?

12. Professor Johnson is asked by an editor at the *Journal of Chemical Phys-
     ics* to review a paper that is closely related to her own work on polymer
     configurations in solution.

     a.   She thinks it would be educational for the students in her research
          group to read the paper. Should she distribute copies to her
          students?

b. What if she sees some references in the new paper that she had not known about? Is it ethical for her to add those references to her own files?

13. You see a newly published paper by your colleague and collaborator and then you are astonished to see that your name is on the paper. You were not consulted about the project and you knew nothing about the paper. What do you do?

14. Frank is an undergraduate student who arranged to do research for credit with Professor Jones. It immediately became clear that Frank was not very interested in the research, but thought that having research on his transcript would be good for his applications to medical school. He was supposed to spend Wednesday and Friday mornings on the research, but he often failed to show up. All in all, he did about eighty hours of work over the semester, rather than the 120 hours for which he had contracted with Jones. When he did collect data, he analyzed it improperly, and his errors had to be corrected by Jones. At the end of the semester, Frank received a "C" for the research course. When Jones writes the work up for publication, should Frank's name be on the paper as a coauthor?

15. Professor Thomas has been juggling three research grants to keep his laboratory running. He is the sole support of his family, including two children and his aged mother. He supports six graduate students and pays his own summer salary. Two students have one year left before finishing their PhD degrees; the other four have several years to go. Suddenly he learns that one of the grants will not be renewed, reducing his resources by 30 percent. What can he do?

Inquiry Questions on the Scientist and Justice

1. Look up the case of astrophysicist Jocelyn Bell. Discuss the ethical issues involved in her mentee/mentor relationship. To what extent are mentors responsible for the scientific products of their mentees? To what extent can mentors take credit for the scientific products of their mentees? What are the complexities caused by the power differentials in such relationships?

2. It is a famous story in organic chemistry that German chemist Friedrich Kekulé proposed the ring structure of benzene in 1865 after dreaming

of a snake biting its tail. It is now known that Josef Loschmidt had proposed ring structures for organic compounds in 1861.[38] Students of organic chemistry do not hear about Loschmidt. Find out more about this issue. Why do you think that Kekulé got all the credit? Is it just? Does it matter?

3. Organic chemist Robert Woodward and physical chemist Roald Hoffman collaborated in 1964 to produce the *Woodward–Hoffman rules* that govern certain organic reactions.

    a.   Jeffrey I. Seeman has written about their collaboration and has suggested the questions listed below:

        • What do you imagine that Woodward and Hoffmann each experienced during the first phase of their collaboration ... ?

        • What do you imagine was the nature and character of their collaboration?

        • What do you imagine was the extent that outside factors, such as the fear that someone else would publish first or Hoffmann's need to secure an academic position, influenced Woodward and Hoffmann's behaviors and the timing of the first Woodward–Hoffmann paper?[39]

    Read Seeman's paper and think about these questions.

    b.   Seeman's paper reports accusations made against Woodward and Hoffman by fellow chemist E. J. Corey. What is your assessment of Corey's accusations?

    c.   Note in Seeman's paper his use of the laboratory notebooks of Hoffman. Compare Hoffman's notebooks to the recommendations on notebooks in chapter 3 and appendix A.

4. In about 2013, a website called *PubPeer* was started that aims to allow anonymous, unrefereed comments on published scientific papers.[40] Have a look at this website. What are the advantages of such an open site for criticism? What are the disadvantages? How does this means of review compare to standard journal peer review and to comments in journals on published articles?

5. Look up the number of PhDs produced each year in the United States in your field. Are there too many PhDs, or not enough? What kinds of jobs have new PhDs found in recent years? What is the unemployment rate? If you are a mentor to graduate students, is this an ethical issue for you? If so, what can you do about it?

6. You are a graduate student and you are supervising an undergraduate student in your laboratory. The student is working with an old mercury thermometer that she found in a drawer. The mercury column is broken, so she tries to fix it by putting the bottom of the thermometer into a Bunsen burner. The heat causes the thermometer to break, and the mercury is spilled onto the bench and the floor. What do you do? What are the procedures for reporting a safety problem at your institution? [Note: Mercury thermometers should be avoided due to the toxicity of mercury. Liquid-in-glass thermometers with broken liquid columns should be repaired by cooling, never by heating.]

7. Recall the case of Jan Hendrik Schön from chapter 3. The question is, "What did his coauthors know and when did they know it?" Did the fraud result from a division of labor (such as crystal preparations and device measurements) in which no one except Schön knew all the parts of the experiments? Should a person who accepts an authorship also accept responsibility for error or fraud, even if that error or fraud was committed by one coauthor and was unknown to other coauthors?

8. Is it possible to accidentally plagiarize someone else's work, perhaps in the course of taking notes for a paper or proposal? Would it be good practice for a scientist to run a plagiarism check on his or her own work? Find out about plagiarism programs that are currently available.

9. At the University of Chicago in 1908–1909, Robert Millikan developed the famous oil-drop experiment to determine the charge on an electron. Louis Begeman and Harvey Fletcher were graduate students working with him on the experiment.[41] Begeman and Millikan had been using water droplets suspended in an electrical field, but the water kept evaporating. Fletcher suggested that oil drops replace the water drops, and success followed. Fletcher then replaced Begeman on the project. Millikan made a deal with Fletcher that Millikan would himself be the sole author on the paper on the charge on the electron; there is no acknowledgment of Begeman in the 1910 paper, but Fletcher is mentioned twice.[42] Fletcher was given sole authorship on another (less significant) paper, at the end of which he thanks Millikan.[43] Millikan won the Nobel Prize in 1923 for his work on the charge of the electron and on the photoelectric effect. Fletcher had a distinguished

career at Western Electric and Bell Telephone Laboratories. Read the papers in the references and think about the ethical issues in this case. What might Millikan have done differently? What might Fletcher have done? What became of Louis Begeman?

10. The 1984 explosion of the Union Carbide chemical plant in Bhopal, India, led to more than 2,500 deaths and was due to failure by managers to maintain proper plant maintenance and safety protocols.[44] Find out more about what happened at Bhopal and why. What is the current situation for that plant and those who were injured?

Further Reading on the Scientist and Justice

ACS Committee on Chemical Safety. *Identifying and Evaluating Hazards in Research Laboratories*. Washington, DC: American Chemical Society, 2015.

Baykoucheva, Svetla. *Managing Scientific Information and Research Data*. New York: Chandos Publishing (Elsevier), 2015.

Biagioli, Mario, and Peter Galison, eds. *Scientific Authorship: Credit and Intellectual Property in Science*. New York: Routledge, 2002.

Committee on Science, Engineering, and Public Policy. *On Being a Scientist: Responsible Conduct in Research*. 3rd ed. Washington, DC: National Academy of Sciences, 2009.

LaFollette, Marcel C. *Stealing into Print: Fraud, Plagiarism, and Misconduct in Scientific Publishing*. Berkeley: University of California Press, 1992.

Meyers, Morton. *Prize Fight: The Race and the Rivalry to Be the First in Science*. New York: Palgrave Macmillan Trade, 2012.

National Research Council. *Safe Science: Promoting a Culture of Safety in Academic Chemical Research*. Washington, DC: National Academy of Sciences, 2014.

Wilson, E. Bright. *An Introduction to Scientific Research*. New York: McGraw-Hill, 1952.

Notes

1. Gilbert Newton Lewis and Merle Randall, *Thermodynamics and the Free Energies of Chemical Substances* (New York: McGraw-Hill, 1923).

2. Gary Taubes, *Bad Science: The Short Life and Weird Times of Cold Fusion* (New York: Random House, 1993).

3. Paul Valéry, "Au Sujet du Cimetière Marin," *La Nouvelle Revu Francaise* 40, no. 1, March (1933): 399–411.

4. Committee on Science, Engineering, and Public Policy, *On Being a Scientist: Responsible Conduct in Research*, 3rd ed. (Washington, DC: National Academy of Sciences, 2009).

5. William Broad and Nicholas Wade, *Betrayers of the Truth: Fraud and Deceit in the Halls of Science* (New York: Simon and Schuster, 1982).

6. Eugenie Samuel Reich, *Plastic Fantastic: How the Biggest Fraud in Physics Shook the Scientific World* (New York: Palgrave MacMillan, 2009).

7. R. A. Alpher, H. A. Bethe, and G. Gamow, "The Origin of the Chemical Elements," *Physical Review* 73, no. 7 (1948): 803–804.

8. Istvan Hargittai, *Judging Edward Teller: A Closer Look at One of the Most Influential Scientists of the Twentieth Century* (New York: Prometheus, 2010).

9. Priya S. Niranjan, Peter B. Yim, Jeffrey G. Forbes, Sandra C. Greer, Jacek Dudowicz, Karl F. Freed, and Jack F. Douglas, "The Polymerization of Actin: Thermodynamics near the Polymerization Line," *Journal of Chemical Physics* 119, no. 7 (2003): 4070–4084.

10. Peter Louis Galison, *Big Science: The Growth of Large-Scale Research* (Stanford: Stanford University Press, 1994).

11. Nancy J. Cooke and Margaret L. Hilton, eds., *Enhancing the Effectiveness of Team Science* (Washington, DC: National Academies Press, 2015).

12. Reich, *Plastic Fantastic.*

13. Tony Rothman, "Lost in Einstein's Shadow," *American Scientist* 94, no. 2 (2006): 112–113.

14. Gerald L. Geison, *The Private Science of Louis Pasteur* (Princeton, NJ: Princeton University Press, 1995).

15. Seymour Mauskopf, "Crystals and Compounds: Molecular Structure and Composition in Nineteenth-Century French Science," *Transactions of the American Philosophical Society* 66 (1976): 1–82; A, Laurent, "Action des Alcalis Chlorés sur la Lumière Polarisée et sur l'Économie Animale," *Comptes Rendus de l'Académie des Sciences* 24 (1847): 220.

16. Albert Einstein, "Ist die Trägheit eines Körpers von Seinem Energieinhalt Abhängig?, *Annalen der Physik* 18 (1905): 639–641.

17. B. Andreas, Y. Azuma, G. Bartl, P. Becker, H. Bettin, M. Borys, I. Busch et al., "Determination of the Avogadro Constant by Counting the Atoms in a ^{28}Si Crystal," *Physical Review Letters* 106, no. 3 (2011): 030801 (4 pages).

18. Marcia Barinaga, "UCSF Case Raises Questions about Grant Idea Ownership," *Science* 277, no. 5331 (1997): 1430–1431.

19. Bruce Alberts, Ralph J. Cicerone, Stephen E. Fienberg, Alexander Kamb, Marcia McNutt, Robert M. Nerem, Randy Schekman et al., "Self-Correction in Science at Work," *Science* 348, no. 6242 (2015): 1420–1422.

20. James R. Wilson, "Responsible Authorship and Peer Review," *Science and Engineering Ethics* 8, no. 2 (2002): 155–174.

21. S. Baykoucheva, *Managing Scientific Information and Research Data* (New York: Chandos Publishing [Elsevier], 2015).

22. Edmund G. Seebauer, "Whistleblowing: Is It Always Obligatory?," *Chemical and Engineering Progress* 100, no. 6 (June 2004): 23–27.

23. C. K. Gunsalus, "How to Blow the Whistle and Still Have a Career Afterwards," *Science and Engineering Ethics* 4, no. 1 (1998): 51–64.

24. Constance Holden, "Whistleblower Woes," *Science* 271, no. 5245 (1996): 35.

25. Peter Poon, "Legal Protections for the Scientific Misconduct Whistleblower," *Journal of Law, Medicine and Ethics* 23, no. 1 (1995): 88–95.

26. Vivian Weil, "Mentoring: Some Ethical Considerations," *Science and Engineering Ethics* 7, no. 4 (2001): 471–482.

27. Toni Feder, "Combatting Professional Isolation through Mutual Mentoring," *Physics Today* 69, no. 4 (2016): 29–30.

28. Deborah C. Fort, ed., *A Hand Up: Women Mentoring Women in Science*, 2nd. ed. (Washington, DC: Association for Women in Science, 1995).

29. Ronald Kessler, "Absent at the Creation," *Washington Post Magazine*, no. 6 (April 1997): 16–31.

30. John Jenkin, *William and Lawrence Bragg, Father and Son: The Most Extraordinary Collaboration in Science* (New York: Oxford University Press, 2011); Andrew Brown, *J. D. Bernal: The Sage of Science* (New York: Oxford University Press, 2007); Georgina Ferry, *Dorothy Hodgkin: A Life* (Cold Spring Harbor, NY: Cold Spring Harbor Laboratory Press, 2000).

31. James D. Watson, *The Double Helix: A Personal Account of the Discovery of the Structure of DNA* (New York: Atheneum Press, 1968).

32. Brenda Maddox, *Rosalind Franklin: The Dark Lady of DNA* (London: HarperCollins, 2003); Anne Sayre, *Rosalind Franklin and DNA* (New York: W. W. Norton and Company, 2000).

33. Ruth E. Fassinger, "Mentoring," in *The Sage Encyclopedia of LGBTQ Studies*, ed. Abbie E. Goldberg (New York: SAGE Publications, 2016).

34. Stephanie J. Bird, "Mentors, Advisors, and Supervisors: Their Role in Teaching Responsible Research Conduct," *Science and Engineering Ethics* 7, no. 4 (2001): 455–468.

35. Jyllian Kemsley, "Lab Death Defense Costs $4.5 Million," *Chemical and Engineering News* 92, no. 44 (2014): 9.

36. Robert K. Merton, "The Matthew Effect in Science, II. Cumulative Advantage and the Symbolism of Intellectual Property," *Isis* 79, no. 4 (1988): 606–623.

37. James Woodward and David Goodstein, "Conduct, Misconduct, and the Structure of Science," *American Scientist* 84, no. 5 (1996): 479–490.

38. Baykoucheva, *Managing Scientific Information*.

39. Jeffrey I. Seeman, "Woodward–Hoffmann's Stereochemistry of Electrocyclic Reactions: From Day 1 to the JACS Receipt Date (May 5, 1964 to November 30, 1964)," *Journal of Organic Chemistry* 80, no. 23 (2015): 11632–11671.

40. Jennifer Cousin-Frankel, "PubPeer Co-Founder Reveals Identity—and New Plans," *Science* 349, no. 6252 (2015): 1036.

41. Michael F. Perry, "Remembering the Oil-Drop Experiment," *Physics Today* 60, no. 5 (2007): 56–60; Harvey Fletcher, "My Work with Millikan on the Oil-Drop Experiment," *Physics Today* 35, no. June 1 (1982): 43–47; David Goodstein, "In Defense of Robert Andrews Millikan," *American Scientist* 89, no. 1 (2001): 54–60.

42. Robert A. Millikan, "The Isolation of an Ion, the Precision Measurement of Its Charge, and the Correction of Stokes's Law," *Science* 32, no. 822 (1910): 436–448.

43. H. Fletcher, "Einige Beträge zur Theorie der Brownschen Bewegung mit Experimentellen Anwendungen," *Physikalische Zeitschrift* 12, no. 1 (1911): 202–208.

44. Themistocles D'Silva, *The Black Box of Bhopal: A Closer Look at the World's Deadliest Industrial Disaster* (Bloomington, IN: Traffor Publishing, 2006).

5 The Scientist and Lives

Underrepresented Groups in Science

Mike: "My mom is a chemist."
Friend: "Are you going to be a chemist when you grow up?"
Mike: "No, you have to be a girl to be a chemist."
—Michael G. Greer, age eight, 1978

The above quote is an example of the danger of drawing conclusions from a very small data set (see chapters 2 and 3), and also an example of how social conditioning happens: if the only chemist you have ever seen is a woman, then you may well assume that you have to be a woman to be a chemist. In fact, in 1978 most chemists and physicists were men, and this is still true. Science and engineering in the United States and in Europe have been done mainly by men, and those men have mostly been Caucasian. Until very recently, women and minorities have not been welcomed into the enterprise of science. Here *minorities* will be used inclusively to denote racial and ethnic minorities, sexual minorities, and people with disabilities.

The lack of women and minorities in science and engineering matters for two reasons. First, science and engineering prosper best when the finest minds are at work and when different kinds of minds with different standpoints and different ways of thinking are addressing the problems at hand. If women and minorities do not participate in science and engineering, then science loses many of the best minds, and loses the diversity of perspectives that would amplify and enrich the advancement of science. For example, in primatology, women investigators have introduced very long-term studies and very different new models of primate social and mating behavior than had been considered by earlier male researchers.[1]

Figure 5.1
The hands of chemist Dorothy Crowfoot Hodgkin (1910–1994), who had rheumatoid arthritis, drawn by sculptor Henry Moore. Crowfoot Hodgkin won the Nobel Prize in Chemistry in 1964 for elucidating the structure of vitamin B_{12}. Hands of Dorothy Crowfoot Hodgkin III 1978 lithograph (CGM 486). With permission of the Tate Museum, London. © The Henry Moore Foundation. All Rights Reserved, DACS (2016), http://www.henry-moore.org.

Second, the quest for justice requires that women and minorities be offered opportunities to participate and to have their contributions acknowledged and celebrated. Scientific careers can provide intellectual satisfaction, social approbation, and financial comfort. These rewards should be possible for anyone willing to work to achieve them. Women and minorities need science and engineering, and science and engineering need women and minorities.

A Brief History of Women and Minorities in Science

In the ancient world, there were a few women scientists.[2] Hypatia (370–415) in Alexandria, Egypt, developed methods of algebra and geometry and invented several scientific instruments, including a distillation apparatus and a hygrometer, before being tortured to death by Christian zealots. In the Middle Ages, science was done by men in religious orders and a few women in religious orders also became involved. Hildegard of Bingen (1098–1179) was a German nun who wrote on a wide range of science topics, from physiology to cosmology.

In Europe in the sixteenth through nineteenth centuries, science was done mostly by wealthy upper-class men, as a hobby.[3] Their daughters, sisters, or wives often served as assistants. Marie Paulze Lavoisier was the wife of Antoine-Laurent Lavoisier, the French chemist credited with discovering the nature of combustion and establishing the conservation of mass in chemical reactions. Madame Lavoisier was the assistant, scribe, and illustrator for her husband.[4] About a century later in Germany, Agnes Wilhemine Louise Pockels, whose brother was a physicist, did her own important work on the surface tension of liquids, in her kitchen, while caring for their sick parents.[5]

In the nineteenth and early twentieth centuries, science began to be a profession that required higher education. In the United States, higher education for women began with coeducation at Oberlin College in 1833, followed by the women's colleges: the Georgia Female Academy in 1836, then Mount Holyoke in 1837.[6] Mount Holyoke was founded by Mary Lyon, who had a strong interest in the emerging field of chemistry and who instituted a tradition of excellence in chemistry at Mount Holyoke.[7] When American graduate schools developed, women were not admitted until about 1890, and not until the 1970s for some institutions.[8] In France and Germany, women were admitted to graduate programs from the late 1800s, and many American women went to Europe to study.[9]

At first, the jobs available for women educated in science were limited to teaching jobs in elementary and secondary schools and in women's colleges. One exception was in astronomy, where women were hired for the tedious work of analyzing photographic images.[10] Women did not begin to appear on science and engineering faculties (outside of the women's colleges) until the late 1970s.[11] For most of the twentieth century, jobs in the U.S. federal government were the best opportunities for American women scientists.[12] In recent years, women have been moving into employment in industry, but still have been underemployed, underpaid, and absent from management roles.[13]

People of color—male and female—have only slowly been included in science and engineering in the United States.[14] *Historically black colleges and universities*, or *HBCU*s, had been established specifically to educate black

Figure 5.2
Marie (1758–1836) and Antoine Lavoisier (1743–1794), the "Mother and Father of Chemistry." Painted by Jacques Louis David in 1788. Image © The Metropolitan Museum of Art, http://www.metmuseum.org. Image source: Art Resource, NY.

Figure 5.3
Agnes W. L. Pockels (1862–1935), researcher on the physics of liquid surfaces, in about 1922. This work is in the public domain in the United States.

Americans, beginning in the latter part of the nineteenth century, and included Cheyney University (1837), Howard University (1867), Spelman College for women (1881), and the Tuskegee Institute (1881). A few predominately white colleges did admit black students—notably Oberlin College, where black students were admitted by about 1835—but most black college students attended HBCUs. Many predominately white colleges and universities remained racially segregated until after the 1954 *Brown v. Board of Education* (of Topeka, Kansas) decision of the U.S. Supreme Court, and after the civil rights movement of the 1950s and 1960s. Minority serving institutions, including HBCUs, high Hispanic enrollment institutions, and tribal American Indian institutions, have seen enrollments in science drop as other institutions opened their doors to racial minorities.[15]

Some scientists of color did surmount those difficulties. Notable black male American physical scientists include Percy Lavon Julian (1899–1975) in organic chemistry, William A. Lester (1937–) in theoretical physical chemistry, and S. James Gates Jr. (1950–) in theoretical physics.[16] Black female scientists are rarer. Shirley Ann Jackson was the first African American woman to earn a PhD from the Massachusetts Institute of Technology (in physics in 1973), chaired the Nuclear Regulatory Commission, and became president of Rensselaer Polytechnic Institute.[17] In 2015, Paula T. Hammond was named chair of the Department of Chemical Engineering at the Massachusetts Institute of Technology. Asian-American scientists have included Chien-Shiung Wu (1912–1997), a particle physicist who experimentally verified the invalidity of the law of conservation of parity in nuclear decay, thereby supporting the theory of Tsung-Dao Lee (1926–)

Figure 5.4
Sylvester James Gates Jr. (1950–), Distinguished University Professor, University System of Maryland Regents Professor, and John S. Toll Professor of Physics at the University of Maryland, College Park. His work in supersymmetry and supergravity won him the National Medal of Science and membership in the National Academy of Sciences. By permission of S. J. Gates Jr.

and Chen Ning Yang (1922–).[18] Lee and Yang won the Nobel Prize in Physics in 1957.

In the latter part of the twentieth century, policies of the United States federal government addressed sexism and racism in the workplace. The Civil Rights Act of 1964 made discrimination against employees on the basis of race or sex illegal, but exempted educational institutions. In 1967, Executive Order 11375 made sex discrimination illegal in programs receiving federal funds and therefore applied to most colleges and universities. In 1972, Title IX of the Equal Employment Opportunity Act made the Civil Rights Act applicable to educational institutions. Incited by the launch of the Russian satellite Sputnik in 1957, the American government has since provided financial support for graduate education in science, and thus has offered a socioeconomic mobility in science that was not possible in other professions, but even this support has not led to the full inclusion of women and minorities.

Sexual minorities—lesbian, gay, bisexual, and transgender (LGBT) persons did not begin to have legal protections in the workplace in the United States until the twenty-first century, and that process has encountered much opposition. At the end of 2016, only twenty states and the District of Columbia had laws prohibiting employment discrimination on the basis of both sexual orientation and gender identity, and five states have no legal protection for either sexual orientation or gender identity.[19] The lack of legal protection and the persistence of overt and covert prejudice and discrimination have hindered the advancement of LGBT people in the scientific workforce (discussion follows).

Persons with disabilities, such as limitations to vision, hearing, or mobility, have found it difficult to become scientists, but career paths have become more open for them for two reasons: laws that require accommodations, and new technologies that enable access.[20] The Americans with Disabilities Act of 1990 (ADA), as amended by the ADA Amendments Act of 2008 (P. L. 110–325), declares that employers cannot discriminate against employees on the basis of disabilities, and that employers must make *reasonable accommodations* to enable disabled employees to function in their jobs. Such accommodations may include alterations in the workplace (e.g., installation of ramps), provision of assistance (e.g., readers or sign language interpreters), or installation of technology (e.g., screen readers and speech-recognition software).

Associations to support underrepresented groups in science began to appear in the 1970s: the Association for Women in Science (AWIS), the National Organization for the Professional Advancement of Black Chemists and Chemical Engineers (NOBCChE), the Society for the Advancement of Chicanos/Hispanics and Native Americans in Science (SACNAS), and the National Society of Black Physicists (NSBP). The National Organization of Gay and Lesbian Scientists and Technical Professionals (NOGLSTP) was established in 1983. At the about same time, the larger societies (American Association for the Advancement of Science, American Chemical Society, American Physical Society) began to have subcommittees to focus on issues of women and racial and ethnic minorities.

Current Status and Trends in the United States

The National Science Foundation reported in 2015 that women constitute about 50 percent of the U.S. population (ages eighteen to sixty-four), and constitute about 30 percent of those employed in science and engineering.[21] Women are highly represented as working professionals in psychology (73 percent), well represented in social sciences (60 percent) and life sciences (48 percent), and less well represented in mathematics and computer science (26 percent), in physics and related sciences—including chemistry (28 percent), and in engineering (14 percent). Indeed, the percentage of bachelor's degrees in computer science awarded to women has been decreasing for a decade.

The NSF Scientists and Engineers Statistical Data System (see NSF website) for 2013 shows that women in chemistry (at all educational levels) made up 26 percent of the workforce in 1993, 33 percent in 2003, and 37 percent in 2013. The participation of women with doctoral degrees in the chemical workforce was lower but increasing: 19 percent in 2003 and 25 percent in 2013. The American Chemical Society reported that its 2015 membership was 31.4 percent female.[22]

Racial minorities—Asian American, African American, Hispanic and Latino American, and Native American, male and female—all together constituted about 36 percent of the 2014 population of the United States (ages eighteen to sixty-four) and about 32 percent of working scientists and engineers.[23] The largest minority group in science and engineering is Asian men at 14 percent. Representation by African American and Hispanic scientists is about 17 percent in psychology and social sciences; about 9 percent in physical sciences, life sciences, mathematics, and computer sciences; and about 11 percent in engineering. There is increasing representation over time, but the change is slow.

Because sexual minorities do not always declare their identities, there are no reliable data on the numbers of LGBT scientists. However, professional organizations in science have finally begun to address LGBT issues. The American Physical Society has declined to hold meetings in states that discriminate against the LGBT community since about 1994, and APS formally established a committee on LGBT issues in 2014.[24] For the American Chemical Society, the organizational diversity statement was altered to include sexual orientation in 1998, gender expression in 2001, and gender identity in 2007.[25] Official LGBT receptions at national ACS meetings began in 2002, and the ACS Subdivision for Gay and Transgender Chemists and Allies was created in 2010, after LGBT chemists were denied a place on the ACS Committee on Minority Affairs. For 2017, the ACS elected as president Allison A. Campbell, who is openly lesbian.[26]

Figure 5.5
Carolyn R. Bertozzi (1966–), biochemist, openly lesbian, and the Anne T. and Robert M. Bass Professor at Stanford University. She is known for her work on the roles of sugars in cells. Photographer Armin Kübelbeck, CC-BY-SA, https://creativecommons .org/licenses/by-sa/3.0/. Wikimedia Commons, https://commons.wikimedia.org/ wiki/File:Carolyn_Bertozzi_IMG_9385.jpg.

Persons with disabilities are now about 12 percent of the U.S. population, about 11 percent of all undergraduates, and about 7 percent of graduate students.[27] Persons with disabilities are attaining doctoral degrees in science and engineering in increasing numbers, from about 250 per year in the United States in 2004, to 350 per year in 2014. The ACS established a committee on "Chemists with Disabilities;" the APS has addressed disabilities issues via taskforces and reports. As mentioned previously, advances in computer technology and robotics are leading to progress in accommodations for disabilities. For example, the invention of the three-dimensional printer allows the fabrication of models of chemical structures that can be studied by blind chemists.[28]

Why Has This Happened?

Why have women and minorities been so slow to enter the science and engineering workforce? There is no single reason, but, instead, a complex set of circumstances and barriers.[29]

1. Overt discrimination. Conscious, deliberate, explicit sexism and racism, and prejudices against gay people and people with disabilities, have been present in society for a long time and still are not eradicated. Such discrimination certainly was not dead when the current cohort of senior women and minority scientists was coming of age and entering the professions in the 1960s and 1970s.

At that time, most colleges and universities in the United States explicitly did not hire women into science faculties. Any women who did get hired were employed as laboratory technicians, not as professors, and often they worked—unpaid and unappreciated—in the laboratories of their husbands.[30] In the 1960s, men at American Physical Society conferences would insert slides of women in bathing suits into their scientific presentations, presumably as attempts at humor. This deplorable practice is now defunct, but there are still conferences where the male organizers invite no women speakers. For example, the draft program of the 2015 International Congress of Quantum Chemistry in Beijing showed twenty-four invitees and five chairs and honorary chairs: all men. Several women in quantum chemistry started an online petition to object to the exclusion of women. Here is a comment on the petition made by one of the conference organizers:

Has anyone determined the number of black/ Hispanic/ Asian/ American Indian/ etc. speakers to ensure there is no "racial inequality"? How about the number of speakers from every country on the planet ... ? How about the height of the speakers? ... What about weight? ... What about age? Hair color? Shoe size? Marital status? Claimed sexual orientation? Eye color? Nose length? Ability to hear? Ability to see? Ability to walk? Ability to talk? ... once CCL [Computational Chemistry List] starts down this path there is no end to the amount of whining and complaining that the list will have to endure.[31]

This venomous diatribe does remind us that there is still work to do to make the chemical and physical sciences welcome to all.

Overt discrimination lingers in the primary and secondary school systems in this country. Racial and ethnic minorities are still likely to be in school systems that do not provide the strong backgrounds in science and mathematics that are needed for careers in science.[32] It has long been realized that classroom environments for girls are less supportive than for boys: boys get more attention, textbooks stereotype scientists as male, boys get more access to equipment (especially computers), and so on.[33]

2. Covert, implicit discrimination. All of us, male and female alike, judge men and women differently, even when we know better and even if we want to do better. We are socialized to do so.[34] If people are asked to judge who is the taller of a man and woman who are standing together, they are likely to judge the man to be taller, even when this is not true.[35] When the very same job application or the same proposal or the same technical article is reviewed with a male name as author, the review is better than if the name of the author is clearly a female.[36] Biologist Ben A. Barres, who transitioned from female to male, is the perfect example of gender discrimination. Barres tells a story of giving a seminar, after which an attendee was heard to say, "Ben Barres gave a great seminar today, but then his work is much better than his sister's."[37]

Valian writes of *gender schemas*: expectations that we all have of women and of men, that we carry around in our minds, vestigial relics of our own social conditioning, impervious to education and effort.[38] These schemas affect our perceptions of human behavior. One element of the gender schemas is that men can do mathematics, science, and engineering, and that women cannot do mathematics, science, and engineering. Similar prejudices—schemas—exist with respect to race, sexual orientation, gender identity, and disability status. It is difficult to rise above these prejudices.

Covert discrimination also includes the very masculine climate that has prevailed in science and engineering.[39] This climate has been one of competition more than cooperation. There has been resistance to the use of inclusive language, such as *he or she* rather than just *he*. Scientific work has been more valued than any other part of life, requiring an almost religious devotion, dawn to dusk, seven days a week. Such a climate can drive women and minorities out of science and engineering (see items 4 and 5).

3. Accumulated microinequities. Unconscious, covert prejudices will lead to "minor" inequities that may seem innocent and innocuous. However, these *microinequities* can have macro-effects, and can produce significant cumulative disadvantages over time.[40]

Figure 5.6 shows how small disadvantages can accumulate into very large disadvantages. Think of these disadvantages as reverse interest or as inflation, eating away at career capital. The graph show three levels of annual disadvantage: 1 percent, 2 percent, and 3 percent. Assume that two scientists start at the same point—for example, both have PhD degrees

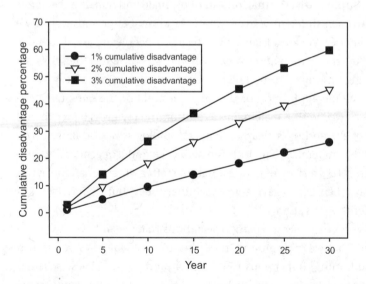

Figure 5.6
The accumulation of disadvantage due to microinequities. If two scientists start at the same level at year 1 and one suffers a 1 percent, 2 percent, or 3 percent disadvantage every year, then that scientist will be at a disadvantage of 26 percent, 45 percent, or 60 percent after a 30-year career.

from distinguished institutions and they are equally smart and equally hardworking. Then even with a 1 percent disadvantage applied every year to one of them, that 1 percent disadvantage accumulates to a 26 percent disadvantage over thirty years of a career.

Such a disadvantage could be 1 percent less lab space, or 1 percent more committee work (so that all committees can have gender and racial representation), or 1 percent more classroom teaching, or 1 percent fewer invitations to speak at conferences, or 1 percent fewer good reviews on papers or proposals. For a scientist with disabilities, this could be 1 percent of his or her time that has to be spent finding accommodations.[41] A disadvantage of 1 percent is hard to discern or measure, but it can have a punishing long-term effect.

Now consider the effects of a 2 percent or 3 percent annual disadvantage, and it is clear that such small annual disadvantages can devastate careers. After thirty years, the female or minority scientist will not be reaping the same rewards as her or his peers, but will be struggling and saddened, exhausted and bitter. Such was the case in the 1990s when women on the faculty of the Massachusetts Institute of Technology went to their administrators with complaints of less laboratory space, less salary, and more service work.[42] Indeed, the fact that women and minorities have to use their time and energy to address these problems is in itself a microinequity, a "shadow job" that adds to their burdens, but for which they receive no compensation or recognition.

Lastly, an important inequity has been the tendency for women and minorities in science to be underappreciated and unrecognized.[43] The prestigious U.S. National Academy of Sciences has about 2,250 members, of which only 13 percent were women in 2013.[44] Until about 2000, very few of the prizes of the American Chemical Society ever went to women. The process for selecting awardees involved secret committees with secret ballots seen only by the committee chairs. When a group of women pressed for an open process, then women began to receive the awards.

4. Differences in career style. Women scientists have been found to have somewhat different styles in pursuing their careers than the styles of men.[45] Women find it harder to establish egalitarian collaborations, which can limit the scope of their work. Women tend to be less self-promoting and less assertive, which can lead to less resources and recognition.[46] Women scientists are more perfectionist—more cautious in making claims from their research data, which means that they write slightly fewer papers. However, the research papers of women are more substantive as indicated by the number of times the papers are cited by other researchers.

Lastly, women scientists tend to avoid the most fashionable research trends and instead find their own niche areas in which to work. Work that is out of the prevailing mainstream will get less attention (fewer citations, fewer invitations to speak), even though such work can lead to great advances. An example of such niche work was the work of astronomer Vera Cooper Rubin. Dr. Rubin set out early in her career to find what she termed "a problem that nobody would bother me about," and her subsequent study of spiral galaxies led to the discovery of dark matter.[47]

5. Work-life conflict. It is not uncommon for scientists to work long hours—sixty or seventy hours a week, over weekends, over holidays. Often the work itself demands attention: an experiment that has to be monitored, a conference that takes a week, a proposal deadline that must be met. Such heavy workloads can discourage potential scientists who want to have a full life as well as a career. Persons with disabilities may need to allow significant time to manage their health and work around their disabilities.

For women, more than for men, there are particular difficulties in managing to have a full life and a demanding profession at the same time. Work-life conflict, or *multiple role management*, is "the most pervasive and persistent challenge facing female science and engineering faculty members—irrespective of the type of institution or the discipline."[48] These problems include not just child care, but also care of friends and elders, because all these responsibilities fall mainly on women.

6. Sexual harassment. Nonconsensual intimate overtures are termed *sexual harassment*. In 2015, a male academic astronomer—Nobel Prize winner Geoffrey Marcy at the University of California, Berkeley—was found guilty of sexually harassing his female students.[49] Marcy resigned his position. Sexual harassment is not limited to astronomers, is not limited to California, and is still happening.[50]

Nonconsensual and unwelcome sexual advances often happen when a person in authority initiates sexual behavior toward a person who has less power. The power dynamics mean that there is no easy way for the person receiving sexual advances to avoid the more powerful perpetrator or to report the offending behavior without suffering career damage, such as poor letters of recommendation or negative reviews of papers and proposals. Sexual harassment undermines the self-respect and self-efficacy of those subjected to it, and contributes to their departure from science.

Sexual harassment is not only crude and rude, and it is not only unethical: it is illegal. Sexual harassment violates Title IX of the 1972 Equal Employment Opportunity Act that prohibits discrimination against women, and violates the hostile workplace laws in many states. Moreover, even consensual intimate relationships between supervisors and supervisees, or between professors and students, are banned by human resource policies in most government entities, businesses, and universities, because of the inherent conflicts of interest (chapter 6). In 2016, the NSF took a stand on its webpage against sexual harassment, and the scientific societies began to establish policies to eradicate sexual harassment in their communities.[51]

The Future for Women and Minorities in Science and Engineering

Just as the low representation of women and minorities in science has no simple cause, the change of that condition has no a simple solution.[52]

First, change must come to the status of women and minorities in society as a whole. This is happening very slowly in the United States. In 2016, on the one hand, the proportion of women is growing in most of the professions, and there are many women in prominent positions in the United States, including three women on the Supreme Court. On the other hand, as of 2016, no woman has ever been elected president of the United States.

Racial and ethnic minorities have seen improvements in their social and economic positions (including the election of an African American president of the United States), but they still suffer deep, insidious, and pervasive prejudice. Sexual minorities have found support and liberation since 2000, including the legalization of same sex marriage and the increased visibility of the transgender population. Nonetheless, there remain deep pockets of rejection of the LGBT community. The differently abled are increasingly being brought into the workplace, but this, too, is incomplete.

The "rate-limiting step" for inclusion in science and engineering may be the rate at which women and minorities choose college majors in science and engineering, which is in part determined by their experiences in primary and secondary education.[53] The American Association of University Women (AAUW) has published a report with specific recommendations for elementary and secondary education, and for college and university environments for both students and faculty members.[54] Among the recommendations are to:

1. Introduce boys and girls to examples of women [and minority] scientists who have succeeded, in order to counter the stereotypes of who can be a scientist;
2. Improve the spatial skills of all children, since such skills are critical to scientific imagination and to manipulation;
3. Encourage all students to take high school mathematics, including the first calculus courses, since mathematics is the gateway to science.

Colleges and universities, too, must continue to remove barriers to the education and career development of women and minorities. Institutional leaders can set the tone, raise awareness, and reward inclusive behavior. In 2006, the administration at a major East Coast state university found that women faculty members in all disciplines (not just science and engineering) were dropping out of the tenure track at a much higher rate than men, and assembled a committee of distinguished senior women faculty members to consider the issue and make recommendations. After much study and discussion, the committee recommended that the most important need for these junior women was good mentoring, and also agreed to organize and provide that mentoring. However, the committee members could not take on this task without some alleviation of their other workloads. The response of the administration was that the committee recommendation was self-serving, and the committee report was discarded without action, thus wasting the time and effort of the members. This story illustrates several points. First, the problems of women were seen as problems to be solved by women (there were no men on the committee), who were to do this shadow job without compensation. Second, the university leadership did not commit itself to solving the problem and withdrew support when resources were required. Third, the committee saw mentoring as a key to change.

Mentoring (see chapter 4) and mutual support from others who share the same experiences are known to increase career success for women and minorities in science and engineering.[55] Mentoring needs to be a part of the work culture and needs to be applied to all, so that women and minorities are not singled out. Moreover, mentoring needs to be monitored and rewarded to assure that it occurs and is effective. One goal of mentoring is to teach women and minorities what to expect from the workplace, and how to access what they need so that the microinequities discussed in this chapter can be eliminated.

Finally, all the causes of the lack of participation of women and minorities in science and engineering listed earlier must be addressed. For women in particular, this means support for work-life management, and eradication of sexual harassment.[56] Work-life conflict can be ameliorated by flexibility with respect to schedule (e.g., flexible tenure timelines), and by supportive benefits (e.g., on-site child care and elder care facilities and parental leave). Most important are a workplace climate and policies that recognize that all people have other beings in their lives who require their time and attention, and that all people have other interests in their lives that enrich them as human beings.

Summary: Underrepresented Groups in Science

In 1960 in the United States, there were 6 percent or 7 percent women working in all science and engineering.[57] In 2015, there were about 30 percent women working in all science and engineering.[58] That is a lot of change in about fifty years. There has been positive change as well for racial/ethnic minorities, sexual minorities, and persons with disabilities. Still, there are many for whom the possibilities of their lives were not realized:

who were told in high school that people like them could not be scientists,
who got discouraged in college and graduate school,
who could not find a job, or who took a low-level job and never escaped,
who worked for free in the laboratories of others,
who taught as adjunct instructors all their lives,
who could not get adequate child care and did not get tenure,
who never got to be chair or dean or department head or division director,
who never won a prize or chaired a conference,
who were harassed for being gay or transgendered,
who could not get accommodations for their disabilities.

We grieve for the unrealized potential of all these women and minorities, and for all the science that did not get done because they were excluded— the questions not asked, the experiments not done, the papers not written, the students not mentored. What has been the cost for science and for humanity? And yet, what benefits there are yet to be reaped when all can be welcomed into science!

Guided Case Study on Underrepresented Groups in Science: Sexual Harassment

You are a female graduate student, just starting in the laboratory of a famous astronomer. You soon see that all the female students are frequently subjected to his sexual advances. Then, on a trip to an observatory, you narrowly avoid such an advance from him. What should you do?

1. Name the values that are involved and consider the conflicts among them. The ethical framework in chapter 1 assigned value to life and to justice. How do those values enter in this case?
2. Would more information be helpful? Is there a university policy about sexual harassment? Do you need to talk to the other women involved? Do you need to talk to someone in the legal office? Can you talk to a mentor?
3. List as many solutions as possible. Your options range from staying and saying nothing, to leaving quietly, to reporting the harassment confidentially, to reporting the harassment but not confidentially.
4. Do any of these possible solutions require still more information? Does a policy require that you first report the issue to an administrator? Can you maintain your own anonymity in this process? What about that of the other women involved? Do you need documentation? Are you willing to let your complaint be known to the professor?
5. Think further about how you would implement these solutions. Can you stay in this research group, whether you report the problem or not? Do you need your own legal representation?
6. Check on the status of the problem. Has anything changed? Have any of the other women reported the problem? Has anything else happened to you?
7. Decide on a course of action. You can hope for institutional support with your course of action.

Discussion Questions on Underrepresented Groups in Science

1. In order to have women and minorities represented on the committees that make hiring and policy decisions, universities usually have at least one female faculty member and one racial/ethnic minority faculty member on each committee. The result is that the few female and minority faculty members get many more committee assignments than do the white male faculty members. How does this connect to the microinequities and shadow jobs discussed earlier? What can be done about it?

2. In 2015, two female scientists submitted a study of gender differences in the experiences of postdoctoral research scientists to a peer-reviewed journal.[59] The anonymous reviewer rejected the article on the grounds that the two women needed to add a male collaborator in order to be sure that their work was not ideologically biased. What is your view of this incident?

Case Studies on Underrepresented Groups in Science

1. You are an African American man who is a scientist in an industrial laboratory and the only person of color in the laboratory. You notice that every Friday evening, the supervisor and the other people in the group go from the lab to a nearby bistro and have a drink together. You have never been invited to join in the outing. You know that work information is exchanged and work agreements are made at these gatherings. What do you do?

2. The chemistry department at a graduate-level university has hired a Latina woman as an assistant professor. She is the first woman and the first Latina in that department. The department chair wants to give her every support to succeed. What actions do you recommend that the chair take, given what you know about mentoring and microinequities? Make a list of suggestions for the chair.

3. You are a physicist in a federal government laboratory. Your research group is convened weekly by your group leader to discuss plans and progress. You are the only woman in the group of thirty. You find it hard to participate in the meeting because the men talk over you and do not listen when you speak. When you make a suggestion, it is ignored until the same suggestion is made later by a man, who gets applauded. You leave every meeting feeling devalued and depressed. What do you do?

4. Professor Smith is a female physics professor at a major state university. She teaches first-semester first-year students in the large 300-student introductory lecture class. She finds that some students are disrespectful: they talk in class, they arrive late and leave early, they address her as "Mrs. Smith" (she is not married and she has a PhD degree), and they do not honor her office hours. Do you think that any of this behavior would be different if she were male? What should she do about it?

5. Professor Anne Jones is chair of a search committee that will recommend the hiring of a new assistant professor of chemical engineering. The search is sharply focused on the area of fluid mechanics, because there is a desperate need for this expertise to balance the department. When the files of the applicants are reviewed, there are no viable female candidates. Dr. Jones cares about women in engineering and is disturbed that there are no female candidates. What do you think she should do?

Inquiry Questions on Underrepresented Groups in Science

1. The discussion in this chapter focused on North American and European scientists. What can you find out about women and minorities in science in China, Japan, India, Africa, or South America?

2. What are the laws in your state with respect to hostile work environments?

3. What is affirmative action and what is the current status of affirmative action in your state?

4. Find out the percentage of scientists with a disability in your discipline. Is that percentage increasing or decreasing? Are there technological advances that can assist persons with disabilities in your field?

5. Find out about the programs at the National Science Foundation, the National Institutes of Health, and the Department of Energy that are designed to increase diversity in science and engineering. Are there data that indicate the level of success of these programs?

6. You are a woman working in an industrial laboratory and you find out that you are pregnant. The laboratory work involves chemicals that are not safe for a fetus. What are the state and federal laws that protect your job security? How do you approach your supervisor about your pregnancy?

7. In 2016, a research study of academic economists showed that extending the timeline for tenure for new parents works to the advantage of men and to the disadvantage of women.[60] Why is this the case? Do you think this result might apply in your discipline? How can this problem be fixed? Are there other cases of policies that are intended to be gender neutral but do not have gender-neutral effects?

8. For the following scientists, find out where and when they were born and educated, what challenges they faced, and what their major contributions to science have been:

Katherine Burr Blodgett

Annie Jump Cannon

Marie Sklodowska Curie

Mildred Spiewak Dresselhaus

Stephen W. Hawking

Paula T. Hammond

Caroline L. Herschel

Dorothy Crowfoot Hodgkin

Grace Murray Hopper

Deborah S. Jin

Maria Goeppert Mayer

Maria Mitchell

Alexandra Navrotsky

Cecelia Payne-Gaposchkin

Ellen Swallow Richards

Vera Cooper Rubin

Johanna M. H. Levelt Sengers

Alan M. Turing

Chien-Shiung Wu

9. You are the head of a government laboratory group and you have just hired a brilliant young theoretical physical chemist who is blind. How do you assure that he has the full support of his colleagues? What kind of online training is available to inform them? What kind of technological support is he likely to need? What reasonable accommodations can you make?

Further Reading on Underrepresented Groups in Science

See also lists of biographies of scientists and of websites in appendix B.

Abir-Am, Pnina G. and Dorinda Outram, eds. *Uneasy Careers and Intimate Lives: Women in Science, 1789–1979*. New Brunswick, CN: Rutgers University Press, 1987.

Alic, Margaret. *Hypatia's Heritage: A History of Women in Science from Antiquity Through the Nineteenth Century*. Boston: Beacon Press, 1986.

Ambrose, Susan A., Kristin L. Dunkle, Barbara B. Lazarus, Indira Nair, and Deborah A. Harkus. *Journeys of Women in Science and Engineering: No Universal Constants*. Philadelphia: Temple University Press, 1997.

Brown, Jeannette. *African American Women Chemists*. New York: Oxford University Press, 2011.

Byers, Nina, and Gary Williams. *Out of the Shadows: Contributions of Twentieth Century Women to Physics*. Cambridge, UK: Cambridge University Press, 2006.

Campbell Jr., George, Ronnie Denes, and Catherine Morrison. *Access Denied: Race, Ethnicity, and the Scientific Enterprise*. New York: Oxford University Press, 2000.

Ceci, Stephen J., and Wendy M. Williams, eds. *Why Aren't More Women in Science? Top Researchers Debate the Evidence*. Washington, DC: American Psychological Association, 2007.

Etzkowitz, Henry, Carol Kemelgor, and Brian Uzzi. *Athena Unbound: The Advancement of Women in Science and Technology*. Cambridge, UK: Cambridge University Press, 2000.

Gornick, Vivian. *Women in Science: Portraits from a World in Transition*. New York: Simon and Schuster, 1983.

Gornick, Vivian. *Women in Science: Then and Now*. New York: Feminist Press, 2009.

Harding, Sandra, ed. *The Racial Economy of Science: Toward a Democratic Future*. Bloomington: Indiana University Press, 1993.

Hargittai, Magdolna. *Women Scientists: Reflections, Challenges, and Breaking Boundaries*. New York: Oxford University Press, 2015.

Holder, Curtis D. "Coping with Class in Science." *Science* 355, no. 6325 (2017): 658.

Hrabowski III, Freeman A. *Holding Fast to Dreams: Empowering Youth from the Civil Rights Crusade to STEM Achievement*. Boston: Beacon Press, 2015.

Jordan, Diann. *Sisters in Science: Conversations with Black Women Scientists on Race, Gender and Their Passion for Science*. West Lafayette, IN: Purdue University Press, 2006.

Kass-Simon, G., and Patricia Farnes, eds. *Women of Science: Righting the Record*. Bloomington: Indiana University Press, 1990.

Malcom, Lindsey E., and Shirley M. Malcom. "The Double Bind: The Next Generation." *Harvard Educational Review* 81, no. 2 (2011): 162–171.

Marzabadi, Cecilia H., Valerie J. Kuck, Susan A. Nolan, and Janine P. Buckner, eds. *Are Women Achieving Equity in Chemistry? Dissolving Disparity and Catalyzing Change, ACS Symposium Series 929*. Washington, DC: American Chemical Society, 2006.

National Research Council. *Beyond Bias and Barriers: Fulfilling the Potential of Women in Academic Science and Engineering*. Washington, DC: National Academy of Sciences Press, 2006.

Pollack, Eileen. *The Only Woman in the Room: Why Science Is Still a Boys' Club*. Boston: Beacon Press, 2015.

Rayner-Canham, Marelene, and Geoffrey Rayner-Canham. *Women in Chemistry: Their Changing Roles From Alchemical Times to the Mid-Twentieth Century*. Washington, DC: American Chemical Society, 1998.

Rosser, Sue V. *Re-Engineering Female Friendly Science*. New York: Teachers College Press, 1997.

Rosser, Sue V. *Breaking into the Lab: Engineering Progress for Women in Science*. New York: New York University Press, 2012.

Rosser, Sue V. *Academic Women in STEM Faculty: Views beyond a Decade after POWRE*. New York: Palgrave Macmillan, 2017.

Rossiter, Margaret W. *Women Scientists in America: Struggles and Strategies to 1940*. Baltimore, MD: Johns Hopkins University Press, 1982.

Rossiter, Margaret W. *Women Scientists in America: Before Affirmative Action 1940–1972*. Vol. 2. Baltimore, MD: Johns Hopkins University Press, 1995.

Rossiter, Margaret W. *Women Scientists in America: Forging a New World since 1972*. Vol. 3. Baltimore, MD: Johns Hopkins University Press, 2012.

Schiebinger, Londa. *The Mind Has No Sex: Women and the Origins of Modern Science*. Cambridge, MA: Harvard University Press, 1989.

Sobel, Dava. *The Glass Universe: How the Ladies of the Harvard Observatory Took the Measure of the Stars*. New York: Viking, 2016.

Swaby, Rachel. *Headstrong: 52 Women Who Changed Science—and the World*. New York: Broadway Books, 2015.

Valian, Virginia. *Why So Slow? The Advancement of Women*. Cambridge, MA: MIT Press, 1999.

Warren, Wini. *Black Women Scientists in the United States*. Bloomington: Indiana University Press, 2000.

Wasserman, Elga. *The Door in the Dream: Conversations with Eminent Women in Science*. Washington, DC: Joseph Henry Press of the National Academy of Sciences, 2000.

Yoder, Jeremy B., and Allison Mattheis. "Queer in STEM: Workplace Experiences Reported in a National Survey of LGBTQA Individuals in Science, Technology, Engineering, and Mathematics Careers." *Journal of Homosexuality* 63, no. 1 (2016): 1–27.

Films on Underrepresented Groups in Science

1. Bowie, Ben, Stephen Finnegan, and Stephen Hawking. *Hawking*. Documentary film directed by Stephen Finnegan. BBC Film, 2004. Internet streaming from Amazon.com, 90 min.

McCarten, Anthony, and Jane Hawking. *The Theory of Everything*. Dramatic film directed by James Marsh, with Eddie Redmayne as Stephen Hawking. Universal Studios, DVD, 124 min.

Physicist Stephen Hawking, who has amyotrophic lateral sclerosis that has left him physically paralyzed, has applied his mind to the great issues of cosmology.

2. *Breakthrough: The Changing Face of Science in America*. Documentary film produced by Joseph Blatt, Andretta Hamilton, and Henry Hampton. Blackside, Inc., PBS Video, 1996. Videocassette (VHS), about five hrs.

This film series tells the stories of twenty men and women of color who are scientists.

3. Curie, Eve, Paul Osborn, and Hans Rameau. *Madame Curie*. Dramatic film directed by Mervyn LeRoy, starring Greer Garson and Walter Pidgeon. MGM, 1943. Warner, DVD, 124 min.

This biography of Marie and Pierre Curie, while quite romanticized, conveys the joy of their scientific work.

4. *Discovering Women*. Produced by NOVA/WGBH Boston. Documentary film directed by David Sutherland. Films for the Humanities and Sciences, 1994. Videocassette (VHS), about six hrs.

This film series explores the lives and careers of six women scientists.

5. Hodges, Andrew, and Hugh Whitemore. *Breaking the Code*. Dramatic film directed by Herbert Wise, with Derek Jacobi as Alan Turing. PBS Video, 1997. Videocassette (VHS), 75 min.

Moore, Graham, and Andrew Hodges. *The Imitation Game*. Dramatic film directed by Morten Tyldum with Benedict Cumberbatch as Alan Turing. Anchor Bay, 2015. DVD, 113 min.

Alan Turing was a British mathematician who solved the German secret code during World War II and who made early contributions to the development of computers. Turing was a homosexual who was persecuted into committing suicide.

Research with Human Participants

(with Ruth E. Fassinger)

The experiment should be such as to yield fruitful results for the good of society, unprocurable by other methods or means of study, and not random and unnecessary in nature.
—The Nuremberg Code[61]

Research with human participants can arise not only in medicine, biological sciences, and psychology, but also in chemistry, biochemistry, physics, and engineering. For example, a mechanical engineer may develop prosthetic devices for human use, a physicist may work on imaging devices for medical diagnosis, or a chemist may develop polymeric materials to use in skin grafts. Physical scientists can be involved in investigations in education, perhaps testing new ways of teaching physics concepts or exploring how to encourage women and minorities in science. When those educational investigations are meant to be published, then that work must abide by human subjects regulations.

The terms *human subjects* and *human participants* are both used to refer to people who participate in scientific research. The latter term especially respects the contributions that people make voluntarily to science. The two terms will be used interchangeably in this chapter.

Recent History of Concern about Research with Human Participants

Concern about research on human participants escalated at the end of World War II, when the shockingly inhumane Nazi experiments on human beings came to light.[62] Nazi doctors had used inmates from the concentration camps, people who had no choice as to whether to participate and who were subjected to painful, disfiguring, debilitating, and lethal procedures. Some of the experiments concerned questions related to the war, such as whether a pilot in a crashing airplane could survive the fall, or whether a pilot downed into the sea could survive the cold water. Some of the experiments had other purposes, such as a cure for homosexuality, or tests of organ transplantation, or the nature of twins. The justifications given later by the doctors were the utilitarian arguments that the experiments saved more lives than they cost, and that those inmates were going to die anyway.[63]

The trials of the German doctors led to the 1949 Nuremberg Code for human subjects research. The Code had ten points. (The 1949 language uses *he* and *him*, and is made more inclusive here.)

1. The voluntary consent of the human subject is absolutely essential. This means that the person involved should have legal capacity to give consent; should be so situated as to be able to exercise free power of choice, without the intervention of any element of force, fraud, deceit, duress, over-reaching, or other ulterior form of constraint or coercion; and should have sufficient knowledge and comprehension of the elements of the subject matter involved, as to enable him [or her] to make an understanding and enlightened decision. ...

2. The experiment should be such as to yield fruitful results for the good of society, unprocurable by other methods or means of study, and not random and unnecessary in nature.

3. The experiment should be so designed and based on the results of animal experimentation and a knowledge of the natural history of the disease or other problem under study, that the anticipated results will justify the performance of the experiment.

4. The experiment should be so conducted as to avoid all unnecessary physical and mental suffering and injury.

5. No experiment should be conducted, where there is an a priori reason to believe that death or disabling injury will occur; except, perhaps, in those experiments where the experimental physicians also serve as subjects.

6. The degree of risk to be taken should never exceed that determined by the humanitarian importance of the problem to be solved by the experiment.

7. Proper preparations should be made and adequate facilities provided to protect the experimental subject against even remote possibilities of injury, disability, or death.

8. The experiment should be conducted only by scientifically qualified persons ...

9. During the course of the experiment, the human subject should be at liberty to bring the experiment to an end, if he [or she] has reached the physical or mental state, where continuation of the experiment seemed to him [or her] to be impossible.

10. During the course of the experiment, the scientist in charge must be prepared to terminate the experiment at any stage, if he [or she] has probable cause to

believe, in the exercise of the good faith, superior skill and careful judgment required of him [or her], that a continuation of the experiment is likely to result in injury, disability, or death to the experimental subject.[64]

The Nuremberg Code contained most of the basic points that are still used in assessing human subjects research. The World Medical Association (WMA) developed the Declaration of Helsinki in 1964 as a code for research with human subjects that is very similar to the Nuremberg Code, and the WMA has continually revised the Declaration since then (see the WMA website).

These codes for human subjects research were *guidelines*, but lacked the force of law in the United States until many years later. The National Research Act of 1974 led to the Belmont Report of 1978 and to the adoption of federal regulations governing human subjects research.[65] In the United States today, the legal code for human subjects research is set by the Office of Human Research Protections (OHRP) of the Department of Health and Human Services, and is referred to as the *Common Rule regulations* (discussion follows). In addition, professions in which the use of human research participants is frequent (e.g., psychology) often have their own codes of ethics that apply the regulations more specifically.

What Is *Research*? What Is a *Human Subject*?

Research means a systematic effort to increase knowledge that is intended for publication.[66] Collections of data that are not analyzed for new knowledge are not considered research. For example, surveys and educational tests are not usually considered research.

A key question in human subjects research is "What is a human being?" At what point is a fertilized human egg considered human? What about the tissues of aborted fetuses? What about the remains of human beings who have died? What about samples of tissue or bodily fluids from living humans? The most distressing examples of research on human beings are cases where subsets of people have been considered to be not human: people who were incarcerated for crimes, or people who were Jewish, or people who were of a particular race, or people who were mentally disabled.

First, a human subject is defined legally as a *living human being*.[67] Regulations about human subjects do not apply to dead bodies or to bodily tissues that are not identifiable. However, the Health Insurance Portability and

Accountability Act (HIPAA) Privacy Rule of 1996 does govern dead bodies and their tissues, and requires that private information about the subjects be carefully respected. For example, HIPAA forbids the use of existing databases or specimens unless the subjects either are anonymous or have given *informed consent* (discussion follows), even if they are no longer living.

The use of fetal tissue and of human embryonic stem cells in research has been a controversial issue, but is legal in the United States as of 2017. If the donors of the tissue are not identifiable, then such projects are not considered human subjects research. Regulations regarding fetal tissue and stem cells change with time, so it is best to refer to the newest information on the OHRP and National Institutes of Health websites.

Guidelines for Research with Human Participants in the United States

The OHRP Common Rule regulations require that each research institution establish a human subjects *Institutional Review Board* (IRB), and that all research with human participants be reviewed by the IRB *before* the research begins. The IRB consists of at least five members, one of whom is not a scientist and one of whom is not from that institution. The research project is reviewed annually while it is in progress. Research institutions routinely apply these rules to all human participants research, regardless of the source of funding. Failure to follow federal rules can cause loss of funding, can jeopardize rights to intellectual property, can prevent publication, and can lead to criminal prosecution.[68]

Some activities may need only brief review by the IRB. For example, research using biological specimens (hair, nail, blood) collected by noninvasive means, or using materials already collected for some other purpose, may not be subject to full review. In any case, the IRB should be consulted.

The IRB considers the following aspects of the proposed research:

1. Informed consent. Human participants must give their written consent to be a part of a research study, and that consent must be *free* and *informed*. *Free* means that the participant must give consent voluntarily, with no implicit or explicit coercion. *Informed* means that the participant must be told of the exact nature of the study and of any risks involved. The participant must be competent to understand the information conveyed and to make a judgment about participation. Thus the participant must be helped to understand any technical language in the consent form. For example, many people may not understand the term *randomization* and its implication that the participant may or may not receive any treatment that is being studied.

Forms for participant consent must be approved by the IRB of the institution and signed by the participant or by the legal guardian of the participant. Sample forms can be found on the Internet. Often the signature of a witness is included, but this is not required. It is best if the participant is given a copy of the form in time for any questions to be answered fully. The form cannot require that the participant forego the right to sue the researcher. The form must reveal any financial interest that the researcher has in the outcome of the research, since that can lead to a conflict of interest for the researcher (chapter 6).

Children are not considered competent to give informed consent. For research with children, the IRB must assess the degree of risk and then decide the nature of consent. Usually consent for children is taken as the consent of one or both legal parents or guardians. Then the ethical issues become the competence of the parents or guardians to understand the project, and the dedication of the parents or guardians to the welfare of the children.

Research involving women who are pregnant or who might become pregnant requires special consideration. The NIH Revitalization Act of 1993 requires the inclusion of women and minorities in research studies (see item 6). However, risks to a fetus must be assessed and plans must be made to monitor the fetus during the research program.

Research with people who are cognitively impaired, whether because of a developmental disorder, an age-related dementia, or a psychiatric disorder, must be handled with careful consideration. The Common Rule disallows people who are in institutions because of their impairments from being subjects for research projects that have no relation to their disorders. The inclusion of such people in research studies that are related to their disorders is allowed, but with careful analysis of the degree to which they or their guardians can give free and informed consent.

Before the advent of the Common Rule, prisoners incarcerated for criminal offenses were often enticed and coerced into serving as human subjects in scientific studies, by being offered special advantages within the prison system, or by being threatened with the loss of advantages if they did not cooperate. For example, from 1951 to 1974, Albert M. Kligman used prisoners—mostly African American men—at Holmsberg Prison in

Philadelphia to study the dermatological effects of cosmetics, shampoos, and medications on behalf of various companies.[69] For the U.S. Central Intelligence Agency and the U.S. Army, Kligman tested psychoactive drugs and chemical warfare agents. Some tests involved injecting live viruses, bacteria, and carcinogens into the prisoners. He did not obtain informed consent from the prisoners; they were ignorant of the nature and purpose of the tests. The prisoners received a few hundred dollars in compensation. Many suffered permanent physical and mental damage. Kligman incurred only a brief debarment from Federal Drug Administration funding and became very wealthy from his Holmsberg work. In the United States today, research with prisoners as subjects is limited to studies of "the cause and effect of incarceration and crime; ... of prisons or of incarcerated persons; ... of conditions that affect prisoners en masse; and therapeutic studies."[70] Prisoners must give informed consent and can be subjected to no more than minimal risk.

A case of grave violation of modern informed consent protocols was that of the infamous Tuskegee syphilis studies.[71] From 1932 to 1972, the Public Health Service of the U.S. government followed the progress of about six hundred African American men with syphilis in Macon County, Georgia, pretending to treat them but actually giving them no treatment at all, and instead using them to learn about the progress of the disease. Over that forty years, many of the men died of syphilis, even though the penicillin that would have cured them was available after 1940. Not only did the men not give informed consent, they were *misinformed* as to the purpose and effect of the treatment. It was the Tuskegee study and its mistreatment of human subjects that led the U.S. Department of Health and Human Services to require the establishment of IRBs.[72]

2. Deception. There may be cases in which the full disclosure of the nature of the study would undermine the very purpose of the study. Then participants have to be deceived and misinformed, and thus their consent cannot be fully informed. For example, if the purpose of a study is to determine whether female physics students are more successful with a female instructor than with a male instructor, then informing the students and instructors of that purpose could affect their behavior and bias the results. In such cases, the experimenter must explain fully to the IRB why deception is necessary and how the welfare of the participants will be protected, must debrief the participants as soon as possible (see item 8), and must be prepared to allow the participants to withdraw from the study even after the data have been collected.

3. Confidentiality. The human participants must be guaranteed that the information collected about them will be kept confidential. This is usually stated within the informed consent form.

4. Minimal impact on participants. The study must be designed to minimize the impact on the participants, including reducing all physical, psychological, social, and legal risks.

A famous study in psychology that resulted in unanticipated negative impact was the Stanford Prison Study by Philip Zimbardo in 1971.[73] Zimbardo and his colleagues wanted to test the effects on human beings of prison conditions, particularly the differential impact on prisoners and prison guards. They assembled a group of white male college students, pretested them to assure their mental health, randomly assigned some to be *guards* and some to be *prisoners*, and installed them in a make-believe prison on the campus of Stanford University. The experiment was intended to last two weeks, but had to be stopped after six days because the guards were inflicting severe psychological trauma on the prisoners, and the prisoners were showing emotional and physical stress. Films of this experiment are available on the Internet. Some of the participants in this study not only suffered unexpected trauma during the experiment, but have reported suffering psychological and social effects for the rest of their lives. This study would not be permitted under modern human participants protocols.

5. Representative sampling. A study using human subjects must include a diverse sample of people that is consistent with their representation in the general population or consistent with the aims of the study, or both, unless a strong justification can be made to do otherwise. The sample of participants should be diverse in terms of gender, race, age, and other factors. A representative sampling is not only a matter of fair treatment for all members of society, but also a necessity for a meaningful experimental outcome (chapter 3).

From its inception in 1887, the National Institutes of Health allowed research projects to exclude women. The justification was that women were too complicated because of hormonal cycles, pregnancy, and menopause. The result was that no research was done on such health concerns as heart

Figure 5.7
Physician Bernadine P. Healy (1944–2011), director of the NIH from 1991 to 1993.
This work is in the public domain in the United States.

disease in women. It was not until Bernadine P. Healy became the first
female director of the NIH in 1991 that this issue was addressed. Healy's
impact included the NIH Revitalization Act of 1993 that requires the inclusion of women and minorities in research studies, and the Women's Health
Initiative, a fifteen-year program aimed at improving research on women's
health.

6. Statistical validity. The scientist who proposes to use human participants must make a priori calculations to assure that the data collected
will suffice to provide a statistically valid conclusion. It is disrespectful of
human life to waste time, energy, and materials on studies that cannot pos-
sibly lead to a convincing result.

7. Value of scientific knowledge sought. The proposed research must
intend to seek new scientific information that is worth having. The proposing scientists are charged with persuading the IRB that the pursuit is
important and that its results will matter.

8. Debriefing of human participants. At the end of the research participation, the human participants must be provided with a summary of the research. Any possible negative effects must be explored and mitigated. Participants must be allowed to withdraw their data if they choose to do so.

The Common Rule regulations have been in place since 1991, but they are under review for minor changes in 2018. Researchers can consult the HHS website for the revised regulations.

Bioethics

The field of bioethics includes not only research with human and animal subjects, but also stem cell research, human cloning, reproductive technology, species sustainability, and genetic modification. These latter topics are not included in this book aimed at chemists and physicists, but references are given in appendix D for the reader who wishes to explore them. Also not discussed here are the special regulations involved in the development of drugs, medical devices, and surgical procedures.

Summary: Research with Human Participants

Research on human participants in the United States is regulated through the OHRP of the Department of Health and Human Services. Individual institutions implement the law through their IRBs. The law requires informed consent from participants, confidentiality of personal information, minimal impact on participants, sampling representative of the general population or the study population or both, prior evidence of the statistical validity and importance of the study, and post-participation debriefing of the participants.

Guided Case Study on Research with Human Participants: Subjects Who Cannot Give Informed Consent

You are a physicist who is developing a new kind of medical imaging technique to assess and analyze brain damage in elderly patients with Alzheimer's disease. The study may lead to a patent and to a commercial product on which you stand to make money. You have your first prototype machine ready to test. You need elderly patients on which to make your tests, and you need subjects with various stages of dementia. A local nursing home has offered to cooperate in your study. How do you proceed?

1. Name the values that are involved and consider the conflicts among them. Think about the value system developed in chapter 1. How do you allow for human autonomy in this case?

2. Would more information be helpful? Review again the human subjects guidelines given above. What is required for informed consent? Will informed consent depend on the extent of dementia? How do you maintain confidentiality? How do you meet the requirement for debriefing?

3. List as many solutions as possible. Design a consent form for your study. Design a method of concealing subject identities during data collection.

4. Do any of these possible solutions require still more information? Who has the power of attorney for each subject? What if some subjects have no person with a power of attorney?

5. Think further about how you would implement these solutions. What if the subjects are frightened by the procedure and refuse to cooperate? Do you need to plan for deception to achieve cooperation?

6. Check on the status of the problem. Has anything changed? Is there a physician, a gerontologist, who could collaborate with you?

7. Decide on a course of action. You will need to be sure that you have enough subjects for the results to be valid.

Discussion Questions on Research with Human Participants

1. Data from the Nazi experiments on human beings still exist and relate to such issues as twin genetics, hypothermia resistance, and organ transplantation. Is it ethical to use these data that were taken under appalling conditions of torture and mutilation?

2. How can a researcher verify that a research participant has fully understood the content of the consent form? How might you assess the level of comprehension? Do you think that more participants will decline to participate if they really understand the studies?

3. Research on chemical and biological weapons (see chapter 6) could require human participants and those participants could be in danger of harm during the studies. Should such studies be permitted? How might participants be recruited and compensated?

4. Consider whether other scholars who use information obtained from people should be required to follow human subjects rules. For example,

what about a historian who is interviewing scientists about the problems they have encountered in their careers?

Case Studies on Research with Human Participants

1. A professor of chemistry teaches a course in research ethics. She is interested in knowing whether such a course actually affects the attitudes and behavior of the students who take the course. She invites a professor of psychology to design a test to be taken by the students, before and after they take the ethics course. They do the tests and the results are very exciting: there is a dramatic effect of the course on the students' attitudes about ethics! Alas, they did not get the permission of the Institutional Review Board before they did the experiment. However, the students' tests are completely anonymous and there is no way of linking a student with a response.

    a. Can they publish the results?

    b. If the students in the class are given extra credit for taking the test on the effect of ethics education, does the extra credit constitute coercion?

2. Chapter 6 will consider the problems of conflicts of interest: ethical dilemmas that arise when one responsibility interferes with another responsibility. An IRB committee can face such conflicts of interest. Consider the following scenarios.

    a. You are on your university IRB and the case before you involves a grant of $5 million to a researcher on the campus. You feel that the research protocol is so weak that it will not lead to useful results. If the IRB denies approval, powerful administrators will be angry. What do you do?

    b. You receive a stipend to your salary for being on the IRB committee. Can this affect your decision making?

    c. The researcher submitting the IRB request is your immediate supervisor. How does this relationship affect your thinking?

    d. The researcher submitting the IRB request is your most detested rival. What would you do?

3. In 1885 in Alsace, a nine-year-old boy named Joseph Meister had been bitten by a rabid dog.[74] A person exposed to rabies has some chance of

not developing the disease, but if the disease appears, the chance of survival is very low. The incubation period between the exposure and the onset of symptoms can be weeks. At that time, Louis Pasteur had developed a vaccine for rabies, but the testing on animals was not yet completed. When Meister's mother appealed to Pasteur, Pasteur agreed to inject the boy with the untested vaccine. Would you have done so? How would modern human subjects protocols affect this decision? [Joseph Meister survived and became a caretaker at the Pasteur Institute in Paris. In 1940, when Nazis wanted to visit the tomb of Pasteur in the Institute, Meister refused to admit them. Depressed by the German occupation of France, Meister committed suicide.]

Inquiry Questions on Research with Human Participants

1. There has been some discussion that the Stanford Prison Study gave misleading results because of bias in the sample of participants. What were the possible sources of this bias and how might they have affected the results?
2. Use the webpage of the Office of Human Research Protections of the Department of Health and Human Services to learn more about the conditions under which informed consent can be waived.
3. Invite the chair or a member of your institutional IRB to speak with your group. What did you learn?
4. Henrietta Lacks was an African American mother of five who died in Baltimore in 1951 at the age of thirty-one, while undergoing treatment for cancer at Johns Hopkins University.[75] Cells from her cancer were taken without her knowledge and were propagated as the ubiquitous *HeLa* cell line, used extensively for research.

    a. At the time of Lacks's death, there was no legal requirement for informed consent. What would be the requirement today?
    b. Some of the research with HeLa cells has resulted in commercial profits. There is still no legal requirement that human participants be compensated for the commercial use of their tissues. Is there an ethical obligation to compensate participants for the use of their tissues?
    c. In 2013, the DNA of the HeLa cell line was sequenced and published, making available very private information about the

descendants of Henrietta Lacks.[76] As of 2015, there was no still legal prohibition against publishing such data. Would such a prohibition inhibit research? What about the issues of consent and privacy? How does this relate to the Health Insurance Portability and Accountability Act (HIPAA)?

5. What if, in spite of all the precautions, a participant of a study is somehow harmed by the study?[77] Who should be responsible for addressing that injury?

6. Critics claim that the use of IRBs to control human participants research has failed, that IRBs lack the expertise to do the job, that they waste resources, and that they impede important research.[78] Learn more about this criticism. What solutions have the critics offered?

7. Learn more about the Tuskegee syphilis study. How was the study finally made public and stopped? Discuss the role of Peter Buxton as a whistleblower in this case (see chapter 4).

Further Reading on Research with Human Participants

Annas, George J., and Michael A. Grodin, eds. *The Nazi Doctors and the Nuremberg Code: Human Rights in Human Experimentation*. New York: Oxford University Press, 1992.

Charrow, Robert P. *Law in the Laboratory: A Guide to the Ethics of Federally Funded Science Research*. Chicago: The University of Chicago Press, 2010.

Klitzman, Robert L. *The Ethics Police? The Struggle to Make Human Research Safe*. New York: Oxford University Press, 2015.

Schneider, Carl E. *The Censor's Hand: The Misregulation of Human-Subject Research*. Cambridge, MA: MIT Press, 2015.

Washington, Harriet A. *Medical Apartheid: The Dark History of Medical Experimentation on Black Americans from Colonial Times to the Present*. New York: Doubleday, 2006.

Research with Animal Subjects

There is no fundamental difference between man and the higher mammals in their mental faculties. ... The difference in mind between man and the higher animals, great as it is, certainly is one of degree and not of kind. The love for all living creatures is the most noble attribute of man. We have seen that the senses and intuitions, the various emotions and faculties, such as love, memory, attention and

curiosity, imitation, reason, etc., of which man boasts, may be found in an incipi-
ent, or even sometimes a well-developed condition, in the lower animals.
—Charles Darwin[79]

Animal liberationists do not separate out the human animal, so there is no rational
basis for saying that a human being has special rights. A rat is a pig is a dog is a boy.
They're all mammals.
—Ingrid Newkirk, founder of People for the Ethical Treatment of Animals[80]

The value system that was developed in chapter 1 assigns value to the universe. Animals are a part of the universe, and animals are used in research.
Biologists use animals to study the physiology and sociology of the animals
themselves. Other scientists use animals in lieu of using human beings—for
example, to study genetics or physiology. Medical scientists use animals to
test possible cures for diseases. Bioengineers use animals as reaction ves-
sels to manufacture biological compounds. Industrial chemists use animals
to test cosmetics and skin/hair products. Biochemists use animal tissue to
extract compounds for study.

Philosophers have debated the place of animals in the world and thus
the respect due to them in terms of their ability to reason, plan, and be
self-conscious, or in terms of their ability to communicate via language, or
in terms of their ability suffer pain and distress.[81] The focus here is on the
ability of animals to feel pain. It is readily agreed that many animals suffer
physical pain, and it has been established that many animals suffer psychological pain when kept from social interaction, or held in confined spaces,
or deprived of stimulation.

The utilitarian argument for the use of animals in research is that it makes
possible the reduction of pain and suffering in human lives. The question
is whether the increase in animal pain and suffering in each particular case
is worth the decrease in human pain and suffering. As in other ethical decision making, there are no simple answers, but the methods developed in
chapter 1 will serve to analyze these questions.

Background of Animal Welfare Issues

Humans have for ages used the bodies of animals in order to learn more
about living beings, and especially more about human bodies. The use
of animals in research began to be a controversial issue in Europe in the

mid-1800s. By 1883, Louis Pasteur faced a public campaign to prevent him from using animals in his search for a vaccine for rabies.[82] In England at about the same time, Charles Darwin was a part of the anti-vivisectionist movement that led to the first laws to protect animals.[83]

Modern thinking on animal welfare began in 1975 when Australian philosopher Peter Singer published his landmark book *Animal Liberation*, in which he argued that animal liberation is like any liberationist social movement that requires "an expansion of our moral horizons."[84] His operating tenets were (1) that pain and suffering are to be avoided; (2) that animals feel pain and suffering; (3) that beings—human and animal—have desires about their lives; (4) that we humans are ethically responsible not only for the pain and suffering that we cause, but also for the pain and suffering that we fail to prevent. He advocated an expansion of the value placed on life to include the lives of animals, based on maximizing the good for all beings. Singer did not argue that animals should never be used

Figure 5.8
Peter Singer (1946–), moral philosopher, in 2009. Photo by Joel Travis Sage, courtesy of Creative Commons.

in research, but rather that such research should be minimized. There are groups that support this point of view, including the Humane Society of the United States.

The animal welfare cause was expanded in 1983 by the publication of *The Case for Animal Rights* by American philosopher Tom Regan.[85] Regan argued that animals have the right to their own existences and thus, from a deontological view, he advocated the total abolition of research on animals. Others who believe that animals should never be used in research have formed organizations such as the Animal Liberation Front (ALF) and People for the Ethical Treatment of Animals (PETA). These organizations arose in part because of cases of gut-wrenching mistreatment of animals by some researchers, including the 1981 case of the monkeys in a laboratory in Silver Spring, Maryland, who were abused in the course of neurological research funded by the NIH.[86] There have been incidents in which animal rights activists have destroyed property, released laboratory animals, and threatened people. The Animal Enterprise Protection Act of 1992 outlaws such disruption of research, and its application has led to large fines and prison sentences.[87]

Contemporary Thinking on Animal Research Subjects

In a prescient 1959 book, W. M. S. Russell and R. L. Burch presented the ideas of the "3 Rs" of animal experimentation: reduction, refinement, and replacement.[88] The term *reduction* means reducing the number of animals used to a minimum. *Refinement* refers to the development of ways of reducing stress on animals used in research. *Replacement* means replacing animal subjects with other means of studying the same issues: for example, human cell cultures, plants or microorganisms, computer models. These ideas formed the basis for the current regulations for research using animals.

Thinking about animal research subjects has some common elements with the thinking about human research subjects presented earlier.[89] The current principles for the ethical use of animals in research are as follows:

1. The research issue for which the animals are used should matter. For example, it is more justifiable to use animals to study the effectiveness of an immunotherapy for cancer than it is to study the effectiveness of a hair conditioner. The lives of animals should not be wasted on insignificant studies.

2. The research study should be designed in such a way that the question posed will actually be answered. The number and species of animal test subjects should be calculated to give a statistically significant result. The lives of animals should not be wasted by using too few to be able to answer the research question.

Often research on animal subjects is meant to substitute for research on human subjects. Thought must then be given as to whether results obtained on animals can be taken as valid for human beings. They are not the same, but are they close enough for the research to be useful? This problem is related to the systematic, subjective, and sampling errors discussed in chapter 3. The more the animals are like humans, the more the results can be transferred, but also the more likely the animals are to have emotions and pains that humans have (see question 1 in the Discussion Questions on Animal Subjects section, to follow).

3. The number of animals used should be the minimum needed for a statistically significant result. Projects should not duplicate previous projects without a compelling reason, so replication must have strong justification. The lives of animals should not be wasted by using more than are needed to answer the research question.

The number of animals used in research (including medical research) in the United States remains about 26 million per year, most of which are rats and mice. There is not yet evidence of a decrease in that total number, but there is a shift away from dogs and cats, and toward more mice, rats, and nonhuman primates (such as monkeys and baboons).[90]

4. Research procedures should be designed to minimize animal pain and suffering, both physical and psychological.

5. As much as possible, animal subjects should be replaced with other techniques, such as mathematical models, cell cultures, or robotic simulators.

Guidelines and Laws about Animal Research Subjects in the United States

The principles listed before are the basis for two American federal laws about animal welfare:[91]

1.   The Laboratory Animal Welfare Act (AWA) was first passed in 1966 and has since been updated regularly. The AWA covers all *warm-blooded* animals and applies to all research regardless of funding source. The AWA

is administered by the U.S. Department of Agriculture, Animal and Plant Health Inspection Service.

2. The Health Research Extension Act of 1985 (HREA) covers all *vertebrates* used in research that is funded by the U.S. Public Health Service, of which the National Institutes of Health (NIH) is a part. HREA is administered by the U.S. Department of Health and Human Services, via the NIH Office of Laboratory Animal Welfare.

Warm-blooded animals (mammals, birds) are all vertebrates, and some cold-blooded animals are vertebrates (reptiles, fish, amphibians), so the HREA is more inclusive than the AWA. All invertebrates (insects, spiders, mollusks, worms) are cold-blooded and are not covered at all by federal animal research regulations.

As of January 2016, the NIH requires that biomedical research using animals pay attention to the gender of the animals.[92] As might be expected, most studies have been done on male animals (mice, in particular) and the results on such issues as pain control are different for male animals than for female animals. Studies that limit the animals used to male animals are in violation of the requirement that the study be designed to provide a significant result.

As in the case for human research subjects, the laws require an IRB for animal research subjects: the Institutional Animal Care and Use Committee (IACUC). The number and kinds of members on the IACUC depend on whether the AWA or the HREA is applicable, but an institution usually will set up one committee that will serve for all animal research within that institution.

NIH contracted with the U.S. National Academy of Sciences to produce a reference book for researchers, the *Guide for the Care and Use of Laboratory Animals*, that is available on the NIH and National Academy Press websites. The *Guide* gives details on personnel training, animal feeding and health care, physical facilities, surgery, and euthanasia. Failure to follow these rules can lead to fines or loss of funding or both.

The Association for Assessment and Accreditation of Laboratory Animal Care International is a private, nonprofit association that provides a formal assessment and accreditation of animal care at an institution, based mainly on the *Guide*. Such accreditation is not required by the federal government, but is sought by many institutions in order to confirm their compliance with federal laws and to simplify funding requests.

Researchers who are involved in research with animals in a country other than the United States and who are using U.S. federal funding are bound by U.S. laws. However, some countries may have laws that are more restrictive than are American laws, and the researchers must know and comply with those local laws. Where there is an inconsistency, the most stringent standard generally applies.

Summary: Research with Animal Subjects

Scientists who use animals for research are required by law to consider the value of the research relative to the value of the lives of the animals, and relative to the level of pain and suffering of the animals. Scientists must aim to reduce the number of animals used, must refine techniques so as to reduce stress on the animals, and must replace animals with other techniques whenever possible.

Guided Case Study on Research with Animal Subjects: Choice of Animal Cells

You are a biophysicist exploring the mechanical properties of cells and the effects of compounds such as cholesterol on those properties.[93] You are writing a proposal to the National Institutes of Health to support your work.

1. Name the values that are involved and consider the conflicts among them. You need to choose what kinds of cells to study. The cells need to provide information that can be useful for human cells. How do the values from chapter 1 enter? Are there considerations from chapter 3 with regard to systematic and sampling errors?

2. Would more information be helpful? Would a literature search help? Are there other researchers whom you can consult?

3. List as many solutions as possible. Can you outline a progression of kinds of cells from simple cells (for example, amoeba) to more complex cells (for example, cultured animal cells)?

4. Do any of these possible solutions require still more information? Where can you obtain the animal cells? Does it matter which animal they come from: cow, pig, dog? What kinds of cells have been used by related studies?

5. Think further about how you would implement these solutions. Can you use cultured human cells at the final stage of your study? What are

the ethical issues of using human cells? (See the earlier section Research with Human Participants.)

6. Check on the status of the problem. Has anything changed? Do you need to consider whether the cells come from male or female animals?

7. Decide on a course of action. Maybe you will use a set of various cells?

Discussion Questions on Research with Animal Subjects

1. The more that animals are like human beings, the more the research done on them can be applied to human beings. For example, baboons are more like humans than are mice, so the effects of a carcinogen on baboons are more like the effects on humans, and the effects of the carcinogen on mice are less like the effects on humans. However, the more the animal is like a human, the more likely it is that the animal has awareness and pain sensations that are close to those of humans. How do we address this quandary in terms of the ethics of animal subjects?

2. Over 99 percent of animal research is done on rodents.[94] Public objection to animal research tends to focus on other mammals: cats, dogs, rabbits, and primates. Why do you think that people tend to make a distinction between rodents on the one hand, and other mammals on the other hand? How do people seem to feel about research on invertebrates such as worms? What does this say about the way we humans value various species? How do we justify such *species-ism*?[95]

Case Studies on Research with Animal Subjects

1. The polymerization and depolymerization of the protein actin are important to the way cells keep their structure and the way they move from place to place. In order to understand this reaction, a physical chemist proposes to measure such properties as the heat taken up upon polymerization, the extent of the polymerization as a function of temperature, and so on.[96] She can make measurements using actin from rabbit muscle tissue, or using actin from yeast cells. The molecular structure of actin (the sequence of amino acids in the protein) does not change much from species to species.

   Which should she choose, rabbits or yeast? Keep in mind the following: much more prior work has been done on rabbit muscle actin than on yeast actin; it is a lot easier to extract and purify usable amounts of actin from rabbit muscle tissue than from yeast cells; rabbits are kept in our society as pets, but are also raised as food. Would it matter whether the rabbit was sacrificed in her laboratory for this purpose, versus sacrificed in another laboratory for another purpose with the unused tissue used for the actin experiments, or versus sacrificed by a company that sells the frozen tissue?

2. You are an American scientist and you are collaborating with a scientist in another country. Your own work is funded by the Department of Agriculture and you are thus subject to U.S. laws with respect to animal subjects. Your collaborator has no such rules in her country. Do you ask her to abide by American rules for her part of the work?

3. You are working on the neurobiochemistry of worms. There are no federal regulations to control research on worms. How do you think about the pain and suffering of these animal subjects? Can you assume that worms feel no pain or suffering?

Inquiry Questions on Research with Animal Subjects

1. You are a member of the IACUC (Institutional Animal Care and Use Committee) for your university. Your university is not accredited by the Association for Assessment and Accreditation of Laboratory Animal Care International. You are asked to chair a subcommittee to determine whether to seek this accreditation. What are the reasons for seeking accreditation? What are the reasons for not seeking accreditation?

2. You are a new graduate student in Dr. Lawson's laboratory. Dr. Lawson explains that the research in his group on hormone biochemistry will require that you work with mice. You will need to monitor their behavior after hormone treatment, and you will need to euthanize and dissect mice. How do you prepare yourself for this work? What do you need to know? How will you find the information that you need? How will you account for the genders of the mice? How will you account for the genders of those handling the mice?[97]

3. Compare and contrast the views of Kant and Bentham (see chapter 1) as applied to the use of human and animal subjects in research. How would each philosopher argue for or against human/animal subjects? Would animal subjects be different from human subjects for either philosopher?

Further Reading on Research with Animal Subjects

Carbone, Larry. *What Animals Want: Expertise and Advocacy in Laboratory Animal Welfare Policy*. New York: Oxford University Press, 2004.

Charrow, Robert P. *Law in the Laboratory: A Guide to the Ethics of Federally Funded Science Research*. Chicago: University of Chicago Press, 2010.

DeGrazia, David. *Taking Animals Seriously: Mental Life and Moral Status*. Cambridge, UK: Cambridge University Press, 1996.

Lents, Nathan H. *Not So Different: Finding Human Nature in Animals*. New York: Columbia University Press, 2016.

National Research Council. *Guide for the Care and Use of Laboratory Animals*. Washington, DC: National Academies Press, 2011.

Orlans, Barbara. *In the Name of Science: Issues in Responsible Animal Experimentation*. New York: Oxford University Press, 1993.

Regan, Tom. *The Case for Animal Rights*. Berkeley: University of California Press, 1983.

Regan, Tom. *Defending Animal Rights*. Champaign: University of Illinois Press, 2001.

Russell, W. M., R. I. Burch, and C. W. Hume. *Principles of Humane Experimental Technique*. London: Universities Federation for Animal Welfare, London, 1992.

Singer, Peter. *Animal Liberation*. New York: New York Review Books, 1975.

Singer, Peter. *Writings on an Ethical Life*. New York: HarperCollins, 2000.

Singer, Peter. *Animal Liberation: The Definitive Classic of the Animal Movement*. New York: Harper Perennial Classics, 2009.

Waldau, Paul. *Animal Rights: What Everyone Needs to Know*. New York: Oxford University Press, 2011.

Waldau, Paul. *Animal Studies: An Introduction*. New York: Oxford University Press, 2013.

Notes

1. Linda Jean Shepard, *Lifting the Veil: The Feminine Face of Science* (Boston: Shambhala Publications, 1993).

2. Margaret Alic, *Hypatia's Heritage: A History of Women in Science from Antiquity through the Nineteenth Century* (Boston: Beacon Press, 1986).

3. Ibid.

4. Roald Hoffmann, "Mme. Lavoisier," *American Scientist* 90, no. 1 (2002): 22–24.

5. Elizabeth M. Derrick, "Agnes Pockels, 1862–1935," *Journal of Chemical Education* 59, no. 12 (1982): 1030–1031.

6. Susan A. Ambrose, Kristin L. Dunkle, Barbara B. Lazarus, Indira Nair, and Deborah A. Harkus, *Journeys of Women in Science and Engineering: No Universal Constants* (Philadelphia: Temple University Press, 1997).

7. Miriam R. Levin, *Defining Women's Scientific Enterprise: Mount Holyoke Faculty and the Rise of American Science* (Hanover, NH: University Press of New England, 2005).

8. Margaret W. Rossiter, *Women Scientists in America: Struggles and Strategies to 1940* (Baltimore, MD: Johns Hopkins University Press, 1982).

9. Marelene F. Rayner-Canham and Geoffrey Rayner-Canham, *Women in Chemistry: Their Changing Roles from Alchemical Times to the Mid-Twentieth Century* (Washington, DC: American Chemical Society, Chemical Heritage Foundation, 1998).

10. Rossiter, *Women Scientists in America: Struggles and Strategies to 1940*.

11. Margaret W. Rossiter, *Women Scientists in America: Forging a New World since 1972*, vol. 3 (Baltimore, MD: Johns Hopkins University Press, 2012).

12. Margaret W. Rossiter, *Women Scientists in America: Before Affirmative Action 1940–1972*, vol. 2 (Baltimore, MD: Johns Hopkins University Press, 1995).

13. Ruth Fassinger, Julie Arseneau, Jill Paquin, Heather Walton, and Judith Giordan, *It's Elemental: Enhancing Career Success for Women in the Chemical Industry: Final Report of Project Enhance* (College Park: University of Maryland and National Science Foundation, 2006).

14. G. Campbell, R. Denes, and C. Morrison, *Access Denied: Race, Ethnicity, and the Scientific Enterprise* (New York: Oxford University Press, 2000).

15. National Science Foundation, *Women, Minorities, and Persons with Disabilities in Science and Engineering: 2015. Special Report NSF 15-311* (Arlington, VA: National Science Foundation, 2015).

16. Ronald E. Mickens, *The African American Presence in Physics* (Atlanta: National Society of Black Physicists, 1999).

17. Wini Warren, *Black Women Scientists in the United States* (Bloomington: Indiana University Press, 1999); Diann Jordan, *Sisters in Science: Conversations with Black Women Scientists about Race, Gender, and Their Passion for Science* (West Lafayette, IN: Purdue University Press, 2006).

18. Tai-Chien Chiang, *Madame Chien-Shiung Wu: The First Lady of Physics Research* (Singapore: World Scientific, 2014).

19. Linda Wang, "A Place at the Bench," *Chemical and Engineering News* 94, no. 41 (2016): 18–20.

20. Constance Holden, "Leveling the Playing Field for Scientists with Disabilities," *Science* 282, no. 5386 (1998): 36–37.

21. National Science Foundation, *Women, Minorities, and Persons with Disabilities in Science and Engineering: 2015*; National Science Foundation, *Women, Minorities, and Persons with Disabilities in Science and Engineering: 2017. Special Report NSF 17-310* (Arlington, VA: National Science Foundation, 2017).

22. Steve Marchant and Clint Marchant, *American Chemical Society ChemCensus 2015* (Washington, DC: American Chemical Society, 2015).

23. National Science Foundation, *Women, Minorities, and Persons with Disabilities in Science and Engineering: 2015*; National Science Foundation, *Women, Minorities, and Persons with Disabilities in Science and Engineering: 2017*.

24. Michael Falk and Elena Long, "Is Physics Open and Accepting for LGBT People?," *APS News* 25, no. 3 (2016): 8; Michael Lucibella, "APS to Study Sexual and Gender Diversity Issues in Physics," *APS News* 23, no. 10 (2014): 4; Ramón S. Barthelemy. "From Grass Roots to Changing Policy: LGBT Advocacy in Physics," *AWIS: Association for Women in Science* 48, no. 4 (2016): 20–23.

25. Linda Wang, "Coming Out in the Chemical Sciences," *Chemical and Engineering News* 89, no. 21 (2011): 41–44.

26. Allison A. Campbell, "Pedaling the Power of Chemistry," *Chemical and Engineering News* 95, no. 1 (2017): 33–35.

27. National Science Foundation, *Women, Minorities, and Persons with Disabilities in Science and Engineering: 2015*.

28. Linda Wang, "Not a Disability, but a Unique Ability," *Chemical and Engineering News* 93, no. 45 (2015): 40–41.

29. Ruth E. Fassinger, "Women in Nontraditional Occupational Fields," in *Encyclopedia of Women and Gender: Sex Similarities and Differences and the Impact of Society on Gender*, ed. Judith Worell (New York: Academic Press, 2001); Ruth E. Fassinger and Penelope A. Aser, "Career Counseling for Women in Science, Technology, Engineering, and Mathematics (STEM) Fields," in *Handbook of Career Counseling for Women*, ed. W. Bruce Walsh and Mary J. Heppner (Mahwah, NJ: Lawrence Erlbaum Associates, 2006); Eileen Pollack, "Can You Spot the Real Outlier?," *New York Times Magazine*, October 6, 2013, 30–46.

30. Rossiter, *Women Scientists in America: Forging a New World since 1972*.

31. Colleen Flaherty, "Chemistry Without Women," *Inside Higher Ed*, February 24, 2014,

https://www.insidehighered.com/news/2014/02/24/female-chemists-protest-all
-male-conference-lineup.

32. Lindsey E. Malcom and Shirley M. Malcom, "The Double Bind: The Next Generation," *Harvard Educational Review* 81, no. 2 (2011): 162–171.

33. Bernice R. Sandler, *The Chilly Classroom Climate: A Guide to Improve the Education of Women* (Washington, DC: National Association for Women in Education, U.S. Department of Education, 1996); Myra Sadker and David Sadker, *Failing at Fairness: How Our Schools Cheat Girls* (New York: Simon and Schuster, 1994).

34. Ernesto Reuben, Paula Sapienza, and Linda Zingales, "How Stereotypes Impair Women's Careers in Science," *Proceedings of the National Academy of Sciences US* 111, no. 12 (2014): 4403–4408.

35. M. Biernat, M. Manis, and T. Nelson, "Stereotypes and Standards of Judgment," *Journal of Personality and Social Psychology* 60, no. 1 (1991): 5–20.

36. Corinne A. Moss-Racusin, John F. Dovidiob, Victoria L. Brescoll, Mark J. Mark J. Graham, and Jo Handelsmann, "Science Faculty's Subtle Gender Biases Favor Male Students," *Proceedings of the National Academy of Sciences US* 109, no. 41 (2012): 16474–16479.

37. Ben A. Barres, "Does Gender Matter?," *Nature* 442, no. 7099 (2006): 133–136.

38. V. Valian, *Why So Slow? The Advancement of Women* (Cambridge, MA: MIT Press, 1998).

39. Sue V. Rosser, *Breaking into the Lab: Engineering Progress for Women in Science* (New York: New York University Press, 2012).

40. Mary Rowe, "The Minutiae of Discrimination: The Need for Support," in *Outsiders on the Inside, Women in Organizations*, ed. Barbara Forisha and Barbara Goldman (Upper Saddle River, NJ: Prentice-Hall, 1981); Mary P. Rowe, "Barriers to Equality: The Power of Subtle Discrimination to Maintain Unequal Opportunity," *Employee Responsibilities and Rights Journal* 3, no. 2 (1990): 153–163; J. Cole and B. Singer, "A Theory of Limited Differences: Explaining the Productivity Puzzle in Science," in *The Outer Circle: Women in the Scientific Community*, ed. H. Zuckerman, J. R. Cole, and J. T. Bruer (New York: W. W. Norton, 1991); Gerhard Sonnert and Gerald Holton, "Career Patterns of Women and Men in the Sciences," *American Scientist* 84, no. 1 (1996): 63–71; G. Sonnert and G. Holton, *Gender Differences in Science Careers: The Project Access Study* (New Brunswick, NJ: Rutgers University Press, 1996).

41. Jesse Shanahan, "Disability Is Not a Disqualification," *Science* 351, no. 6271 (2016): 418.

42. Andrew Lawler, "Tenured Women Battle to Make It Less Lonely at the Top," *Science* 286, no. 5443 (1999): 1272–1278.

43. Margaret W. Rossiter, "The Matthew Matilda Effect in Science," *Social Studies of Science* 23, no. 2 (1993): 325–341.

44. Elizabeth Gibney, "Women Under-represented in World's Science Academies," *Nature News & Comment*, February 29, 2016, doi:10.1038/nature.2016.19465.

45. Sonnert and Holton, "Career Patterns of Women and Men in the Sciences."

46. Linda Babcock and Sara Laschever, *Women Don't Ask: Negotiation and the Gender Divide* (Princeton, NJ: Princeton University Press, 2003).

47. Dennis Overbye, "Vera Rubin, Scientist Who Opened Doors for Physics and for Women, Dies at 88," *New York Times,* December 28, 2016, A18.

48. Jane Zimmer Daniels, "The Clare Boothe Luce Program for Women in the Sciences and Engineering," in *Are Women Achieving Equity in Chemistry? Dissolving Disparity and Catalyzing Change*, ed. Cecilia H. Marzabadi, Valerie J. Kuck, Susan A. Nolan, and Janine P. Buckner (Washington, DC: American Chemical Society, 2006).

49. Daniel Clery, "Shining a Light on Sexual Harassment in Astronomy," *Science* 350, no. 6259 (2015): 364–365.

50. Pat Shipman, "Taking the Long View on Sexism in Science," *American Scientist* 103, no. 6 (2015): 392–394.

51. Marcia McNutt, "Societies Can Combat Harassment," *Science* 351, no. 6275 (2016): 791.

52. Beth Mitchneck, Jessi L. Smith, and Melissa Latimer, "A Recipe for Change: Creating a More Inclusive Academy," *Science* 352, no. 6282 (2016): 148–149.

53. Allison K. Shaw and Daniel E. Stanton, "Leaks in the Pipeline: Separating Demographic Inertia from Ongoing Gender Differences in Academia," *Proceedings of the Royal Society B: Biological Sciences* 279, no. 1743 (2012): 3736–3741.

54. Catherine Hill, Christianne Corbett, and Andresse St. Rose, *Why So Few? Women in Science, Technology, Engineering, and Mathematics* (Washington, DC: American Association of University Women, 2010).

55. Deborah C. Fort, ed. *A Hand Up: Women Mentoring Women in Science*, 2nd ed. (Washington, DC: Association for Women in Science, 1995).

56. Deirdre Lockwood, "Balancing the Work-Life Equation," *Chemical and Engineering News* 94, no. 13 (2016): 22–23.

57. Rossiter, *Women Scientists in America: Before Affirmative Action 1940–1972.*

58. National Science Foundation, *Women, Minorities, and Persons with Disabilities in Science and Engineering: 2015.*

59. Rachel Bernstein, "PLOS ONE Ousts Reviewer, Editor after Sexist Peer-Review Storm," *Science*, ScienceInsider, May 1, 2015. DOI: 10.1126/science.aab2568.

60. Justin Wolfers, "Hobbling Women by Trying to Help," *New York Times*, June 26, 2016, BU5.

61. International Military Tribunal, *Trials of War Criminals before the Nuremberg Military Tribunals under Control Council Law No. 10: Nuremberg, October 1946–April, 1949*, vol. 2 (Washington, DC: U.S. Government Printing Office, 1949), 181–182.

62. P. J. Weindling, *Nazi Medicine and the Nuremberg Trials: From Medical War Crimes to Informed Consent* (New York: Palgrave Macmillan, 2005); Robert L. Klitzman, *The Ethics Police?: The Struggle to Make Human Research Safe* (New York: Oxford University Press, 2015); Francis R. Nicosia and Jonathan Huener, *Medicine and Medical Ethics in Nazi Germany: Origins, Practice, Legacies*, Vermont Studies on Nazi Germany and the Holocaust (New York: Berghahn Books, 2002); Robert Jay Lifton, *The Nazi Doctors: Medical Killing and the Psychology of Genocide* (New York: Basic Books, 1986).

63. Marcia Angell, "Medical Research: The Dangers to Human Subjects," *New York Review of Books* 62, no. 18 (2015): 48–51; Marcia Angell, "Medical Research on Humans: Making It Ethical," *New York Review of Books* 62, no. 19 (2015): 30–32.

64. International Military Tribunal, *Trials of War Criminals*.

65. Ibid.; Harriet A. Washington, *Medical Apartheid: The Dark History of Medical Experimentation on Black Americans from Colonial Times to the Present* (New York: Doubleday, 2006).

66. Nicholas H. Steneck, *Introduction to the Responsible Conduct of Research* (Washington, DC: Office of Research Integrity, U.S. Department of Health and Human Services, 2004).

67. Robert P. Charrow, *Law in the Laboratory: A Guide to the Ethics of Federally Funded Research* (Chicago: University of Chicago Press, 2010).

68. Ibid.

69. Allen A. Hornblum, *Acres of Skin: Human Experiments at Holmsburg Prison* (New York: Routledge, 1998).

70. Washington, *Medical Apartheid*.

71. James H. Jones, *Bad Blood: The Tuskegee Syphilis Experiment, New and Expanded Edition* (New York: Free Press, 1993).

72. Charrow, *Law in the Laboratory*.

73. Philip G. Zimbardo, *The Lucifer Effect: Understanding How Good People Turn Evil* (New York: Random House, 2007).

74. Patrice Debré, *Louis Pasteur*, trans. Elborg Forster (Baltimore, MD: Johns Hopkins University Press, 1998).

75. Rebecca Skloot, *The Immortal Life of Henrietta Lacks* (New York: Crown Publishing Group of Random House, 2010).

76. Andrew Adey, Joshua N. Burton, Jacob O. Kitzman, Joseph B. Hiatt, Alexandra P. Lewis, Beth K. Martin, Ruolan Qiu, Choili Lee, and Jay Shendure, "The Haplotype-Resolved Genome and Epigenome of the Aneuploid HeLa Cancer Cell Line," *Nature* 500, no. 7461 (2013): 207–211.

77. Rebecca Dresser, "Aligning Regulations and Ethics in Human Research," *Science* 337, no. 6094 (2012): 527–528.

78. Carl E. Schneider, *The Censor's Hand: The Misregulation of Human-Subject Research* (Cambridge, MA: MIT Press, 2015); I. Glenn Cohen and Holly Fernandez Lynch, eds., *Human Subjects Research Regulation: Perspectives on the Future* (Cambridge, MA: MIT Press, 2014).

79. Charles Darwin, *The Descent of Man, and Selection in Relation to Sex* (London: John Murray, 1871).

80. Adrian Morrison, "Making Choices in the Laboratory," in *Why Experimentation Matters: The Use of Animals in Medical Research*, ed. Ellen Frankel Paul and Jeffrey Paul (Piscataway, NJ: Transaction Publishers, 2001).

81. Martin Benjamin, "Ethics and Animal Consciousness," in *Social Ethics*, ed. Thomas A. Mappes and Jane S. Zembaty (New York: McGraw-Hill, 1982).

82. Dubré, *Louis Pasteur*.

83. Paul Waldau, *Animal Rights: What Everyone Needs to Know* (New York: Oxford University Press, 2011).

84. Peter Singer, *Animal Liberation* (New York: New York Review Books, 1975).

85. Tom Regan, *The Case for Animal Rights* (Berkeley: University of California Press, 1983).

86. Kathy Snow Guillermo, *Monkey Business: The Disturbing Case That Launched the Animal Rights Movement* (Washington, DC: National Press Books, 1993); Caroline Fraser, "The Raid at Silver Spring," *The New Yorker*, April 19, 1993, 66–84.

87. Charrow, *Law in the Laboratory*.

88. W. M. S. Russell and R. L. Burch, *The Principles of Humane Experimental Technique* (London: Methuen, 1959).

89. Z. Bankowski and N. Howard-Jones, *International Guiding Principles for Biomedical Research Involving Animals* (Geneva: Council for International Organizations of Medical Sciences [CIOMS], 1985).

90. Susan Gilbert, Gregory E. Kaebnick, and Thomas H. Murray, *Animal Research Ethics: Evolving Views and Practices: Special Report*, vol. 42 (Garrison, NY: Hastings Center, 2012).

91. Charrow, *Law in the Laboratory*.

92. Editorial, "Why Science Needs Female Mice," *New York Times*, July 19, 2015, SR10.

93. Fitzroy J. Byfield, Helim Aranda-Espinoza, Victor G. Romanenko, George H. Rothblat, and Irena Levitan, "Cholesterol Depletion Increases Membrane Stiffness of Aortic Endothelial Cells," *Biophysical Journal* 87, no. 5 (2004): 3336–3343.

94. Larry Carbone, *What Animals Want: Expertise and Advocacy in Laboratory Animal Welfare Policy* (New York: Oxford University Press, 2004).

95. Richard D. Ryder, "Experiments on Animals," in *Animals, Men and Morals*, ed. Stanley Godlovitch, Roslind Godlovitch, and John Harris (London: Victor Gollanz, 1971).

96. Priya S. Niranjan, Peter B. Yim, Jeffrey G. Forbes, Sandra C. Greer, Jacek Dudowicz, Karl F. Freed, and Jack F. Douglas, "The Polymerization of Actin: Thermodynamics near the Polymerization Line," *Journal of Chemical Physics* 119, no. 7 (2003): 4070–4084.

97. David Grimm, "Male Scent May Compromise Biomedical Studies," *Science* 344, no. 6183 (2014): 461.

6 The Scientist and Society

Science and Public Policy

There is nothing which can better deserve your patronage, than the promotion of Science and Literature. Knowledge is in every country the surest basis of publik happiness.
—President George Washington, State of the Union Address, 1790

The prospect of domination of the nation's scholars by Federal employment, project allocation, and the power of money is ever present and is gravely to be regarded. Yet in holding scientific discovery in respect, as we should, we must also be alert to the equal and opposite danger that public policy could itself become the captive of a scientific-technological elite.
—President Dwight D. Eisenhower, Farewell Address to the Nation, 1961

In 1990, the U.S. Congress agreed to fund the construction of the Super-conducting Supercollider (SSC) in Waxahachie, Texas, at an estimated cost of $4 billion.[1] The SSC was to have been the most advanced facility for elementary particle physics in the world, and would have put the United States in first place for advancing the understanding the nature of matter. The Higgs boson would have been discovered with the SSC, sooner than and instead of at the European Large Hadron Collider. However, after $2 billion had been invested and gigantic tunnels drilled into the Texas soil, in 1993 Congress canceled the SSC.

The SSC was canceled because of poor management that led to cost overruns projected at $10 billion or more. In the end, it was deemed that the projected scientific benefits would not justify the enormous cost.[2] The

Figure 6.1
Excavating for the Superconducting Supercollider, about 1992. Courtesy of SSC,
Fermi National Accelerator Laboratory.

death of the SSC raised a number of questions about how society funds
science:

Who decides which scientific projects should be funded by the government?
What are the criteria for the public funding of science?
How are scientific projects to be monitored and managed?
When should the United States share costs with other countries?
What if the scientific project is politically controversial?
What is the responsibility of scientists to explain their work to the public?

The U.S. government has developed an interest in funding science, both
scientific research aimed at increasing fundamental understanding of the

universe, and scientific research aimed at solving particular societal problems. The public funding of science is an *ethical* issue, related not only to truth as a goal of science, but also to justice in the use of public resources for the good of the public.

A Brief History of the Public Funding of Science in the United States

Until the middle of the twentieth century, science was done with funding from corporate laboratories, from universities, from the scientists' own pockets, or within a few government laboratories. For example, polymer chemist Wallace Carothers developed nylon and neoprene in the 1930s while working at E. I. du Pont de Nemours and Company.[3] Albert A. Michelson's early work (1878) on the speed of light was supported by $2000 from his father-in-law.[4] The National Bureau of Standards, now the National Institute of Standards and Technology, was founded in 1901, had its own laboratories for maintaining measurement standards and reference materials.

Then World Wars I and II brought the government into funding science for weapons, and those projects were the beginning of the collaboration between the government and the community of science in the United States (see the section on Science and Weapons, this chapter). During World War II, the U.S. Office of Scientific Research and Development (OSRD) managed science and engineering contracts for the federal government.[5] Electrical engineer Vannevar Bush was the director of OSRD. At the end of World War II in 1945, Bush presented a plan for continuing public support of basic scientific research. The Bush plan was based on the assumption that a significant portion of that research would, inevitably and directly, lead to applications of value to society.[6] His plan proposed a single independent science agency that would encompass all basic research, including military and medical research.

The Bush plan was not implemented.[7] Instead, a more complex plan was developed by White House advisor John R. Steelman.[8] The National Science Foundation was established in 1950 to support nonmedical fields of research, including physical, biological, and (eventually) engineering and social sciences, by means of grants to external scientists. The National Institutes of Health, dating from 1930, was expanded to support medical and biomedical sciences, both through an internal staff of scientists and through grants to external scientists. Now many other government agencies also support sciences: space science (National Aeronautics and Space Administration, NASA), environmental science (Environmental Protection Agency, EPA), defense research (Office of Naval Research, ONR), energy research (Department of Energy, DOE), and so on.

Support for research by the federal government has *increased* fairly steadily since World War II. Figure 6.2 shows this increase in constant 2000 dollars.[9] There are two "bumps" in the overall upward trend. The first bump begins in the 1950s and corresponds to the launch of the Soviet satellite Sputnik in 1957 and the response of the United States to catch up to the Russians in science and technology. The second bump is related to the decrease in defense funding in the 1980s that followed the end of the Cold War with Russia.[10]

However, the ratio of federal funding for research to the gross domestic product (GDP) has *decreased* over time: from 1.6 percent in 1960 to 0.8 percent in 2010.[11] Support from American industry has increased, as shown in figure 6.2, so that total support (federal and industrial) as a percentage of GDP has held steady at about 3 percent. The U.S. ranks eighth among all countries in the percentage of GDP devoted to all research, below Finland, Sweden, Israel, Japan, South Korea, Denmark, and Singapore.

Industrial research tends to be applied and mission-driven. Bell Laboratories, now owned by Alcatel-Lucent, was once a powerhouse of applied research, where the transistor (1947) and the laser (1953) were invented, but has become less productive in recent decades. The pharmaceutical and

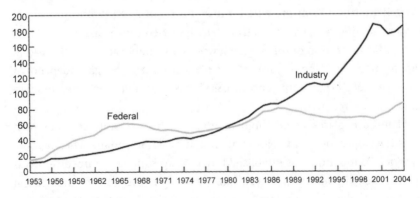

Figure 6.2
Federal and industrial support for research, 1953–2004, in billions of 2000 dollars. This work is in the public domain in the United States.

chemical companies have their own research operations. When industries collaborate with universities, differences of mission can lead to conflicts of interests (addressed later in this chapter).

The support of the U.S. government for scientific research has led to the establishment of the most outstanding research infrastructure in the world. Young scientists from other countries flock to the United States for their graduate and postdoctoral studies. At the same time, government support has led to a dependence of research universities on federal grant funds (see chapter 2). When the federal government funds a research proposal, that funding includes both the money needed directly to pay for the research (for materials, salaries, etc.), and also the *overhead* money or *indirect costs* (30 percent or more of direct costs) that are meant to help pay for the buildings, utilities, library, and so on. The overhead funds have become a significant means of support for university operations. One result is that a faculty scientist in an American research university may be rewarded more for the amount of grant funding that she produces than for her efforts at teaching and mentoring undergraduate and graduate students.

Another result of federal grant funding is that those funds tend to be concentrated in a few big research institutions and in a few states. This raises questions in the U.S. Congress about lack of geographical distribution of grant funds. In 1978, the NSF responded to these political pressures with the Experimental Program to Stimulate Competitive Research (EPSCoR). The 2015 NSF website stated that the "mission of EPSCoR is to assist the National Science Foundation in its statutory function 'to strengthen research and education in science and engineering throughout the United States and to avoid undue concentration of such research and education.'" Researchers in twenty-eight states are eligible to apply for these special funds. In 2015, the total NSF budget was $7.344 billion and the EPSCoR budget was $159.7 million (or 2.2 percent of the total budget). The Department of Energy and other agencies have adopted similar EPSCoR programs.

Federal grant funds are generally distributed after rigorous review by scientific peers (chapters 2 and 3), with scientific merit as the main criterion. EPScOR is an example of research funds that are awarded for which the first criterion is not scientific merit, but for which there is still peer review. There is different issue of *pork barrel* and *earmark* science, in which projects are funded directly by Congress, for which scientific merit is not the first

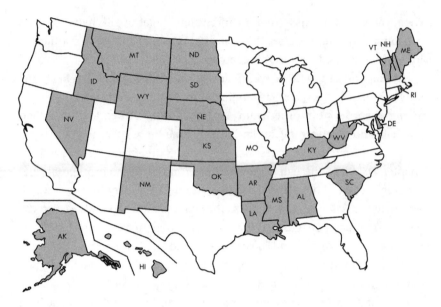

Figure 6.3
States eligible for EPSCoR grants. Guam, Puerto Rico, and the U.S. Virgin Islands are also eligible. Adapted from the United States Department of Energy website, November 2016. This work is in the public domain in the United States.

criterion and for which there is no peer review. Scientific facilities have economic impacts in the places where they are located, so influential members of Congress sometimes seek to protect the interests of their own states—and thus promote their own reelections. The location of the SSC in Texas was a political victory for then Speaker of the House James C. Wright (D-TX).[12]

In 1979, Tufts University in Medford, Massachusetts, hired lobbyists to influence House Speaker Thomas O'Neill (D-MA) to appropriate funds for the university's Jean Mayer USDA Human Nutrition Research Center on Aging.[13] Funds were directed through the U.S. Department of Agriculture (USDA) for the building and the staff; by 2015 the center was still about 45 percent USDA-funded. In 1983, the Catholic University of America in Washington, DC, hired the same lobbyists to get money for a laboratory— the Vitreous State Laboratory—focused on the science and technology of glassy materials. Columbia University did the same to get money for a chemistry building.

In 1983, the American Association of Universities passed a resolution that research funding should be based on "the informed peer judgments of other scientists," not on pork barrel politics.[14] There have been measures taken since 2006 to reduce earmark spending by making the process more open; both houses of Congress now require public disclosure of earmark funding.

Basic Science and Applied Science

The premise of public support of research in science is that the results of the research will help to solve public problems. The question that arises is how closely the research should to be connected to a national need. Science is often divided into *basic* science and *applied* science. Basic or fundamental science is intended to expand our understanding of the world. Applied science is intended to solve specific societal problems. However, such a simplistic dichotomy does not at all represent the complex, interactive web of relationships among basic science, applied science, technology, and engineering.[15]

Basic science has certainly led to many practical applications. Michael Faraday's work on the principles of electricity and magnetism has led, over time, to cell phones and the Internet. The discovery of nuclear fission led to the atomic bomb and to nuclear reactors for electrical energy. Understanding the structure of DNA led to DNA matching for criminal identification. Analytical chemistry is essential to detecting lead in public water systems. Likewise, applied research and technology and engineering have enabled advances in basic knowledge. New technologies aid basic science by providing new ways of extending observations: telescopes and microscopes, computers, instruments such as spectrometers. Applied research can overlap with basic research, as in the case of the Louis Pasteur's exploration of the microorganisms involved in infectious diseases. Even the distinction between science and engineering is not a clear one: engineering researchers today do work that crosses over into fundamental physics, chemistry, and biology. For example, the development of new kinds of batteries requires basic research in electrochemistry.

In 1994, Senator Barbara Mikulski (D-MD) argued that the public interest is better served if basic research (funded by the National Science Foundation, for example) is selected to be in strategic directions. Such strategic research can still be aimed at fundamental understanding, but that new understanding should concern phenomena that are related very directly to particular societal problems. Mikulski said,

Perhaps it's time for NSF to reorganize—over time—into a series of institutes like manufacturing, climate change, high performance computing, or other strategic areas ...

In short, we need to rekindle in the scientific community a new sense of patriotism. That their work is funded by ordinary taxpayers—the checkout clerk at the grocery store or a machinist on the assembly line at GM. It is not an entitlement, and it is not always guaranteed.[16]

The NIH has always been organized into institutes focused on particular health issues (such as the National Cancer Institute, or the National Institute on Aging). In spite of Mikulski's views, the NSF has not reorganized into institutes based on special applied needs, but is still organized around scientific disciplines. However, members of Congress will continue to raise questions about the selection of publicly funded research projects.

The critical issues of public policy for science are who decides which strategic directions are most important for national needs, and which research shows the most promise for success in those directions. These questions get addressed in different ways by different government agencies, and this pluralism is a strength of the American scientific infrastructure. There are the *macroallocations* made by Congress when it sets the budgets for various agencies such as NSF and NIH, and there are the *microallocations* made within an agency.[17] The macroallocations reflect the view of Congress on the problems facing society. The microallocations are judgments of which proposed research will help to solve the problems, based on the opinions of experts within the agencies and the peer reviewers within the scientific community (see chapters 2 and 4). For example, NSF funds research proposals based on two criteria:

Intellectual Merit: The Intellectual Merit criterion encompasses the potential to advance knowledge; and

Broader Impacts: The Broader Impacts criterion encompasses the potential to benefit society and contribute to the achievement of specific, desired societal outcomes.[18]

The competitions at NSF and NIH for microallocations are so fierce that only about 20 percent of the proposals submitted are funded.

The support of research and the resulting development of both understanding and applications builds capacity for more understanding and more applications. Research done in universities with the involvement of undergraduate and graduate students enhances that capacity by developing a workforce of scientists and engineers. Society will continue to debate the ways and means of supporting science, but there is no doubt that such support has been a good investment for the public.

Big Science and Little Science

You can visit the Deutsches Museum in Munich, Germany, and see the very simple apparatus used from 1937 to 1938 by Otto Hahn, Lise Meitner, Fritz Strassmann, and Otto Frisch to discover nuclear fission. With just the table of equipment shown in figure 6.4, they found a profound insight into the nature of matter. Compare this to the SSC discussed earlier and shown in figure 6.1. As nuclear physics developed, larger and larger pieces of equipment were needed, such as nuclear reactors and particle accelerators. These larger facilities then required larger and larger teams of scientists, engineers, and technicians. Nuclear physics began as *little science*—science done by one or two investigators with fairly inexpensive equipment, and became *big science*—science done by large teams and requiring huge investments.[19]

The balance of little science and big science, like the balance of basic science and applied science, is another issue of national science policy. Much

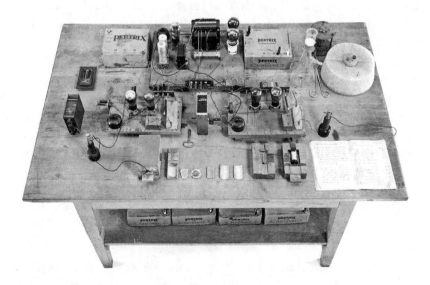

Figure 6.4
Apparatus used for the discovery of nuclear fission. By permission of the Deutsches Museum, München, Archiv, BN24876.

scientific research is still little science, requiring investments of $200,000 to $1 million per year, and such small-scale science is a permanent part of the American national research portfolio. Examples are the standard three-year single investigator grants issued by NSF. There is now also *medium* science: cases where an agency funds a center to bring together scientists from various disciplines to address a particular issue, with a budget of about $5 to 6 million per year. An example is the National Socio-Environmental Synthesis Center (SESYNC) at the University of Maryland, funded by NSF in 2011 to address socioenvironmental problems with interdisciplinary collaboration among physical, biological, and social scientists.

Then there is truly big science, requiring facilities costing billions of dollars. Examples are the Hubble Space Telescope and the Curiosity Mars Rover. The mechanisms for establishing these facilities vary. The Hubble Telescope was funded by the National Aeronautics and Space Administration (NASA) in cooperation with the European Space Agency, and is administered through Johns Hopkins University. The Curiosity Rover is operated by NASA, but the many cameras and analytical instruments aboard Curiosity were devised and built by various companies and by agencies in various countries. Nuclear reactors for research have been built and managed at government sites (e.g., at the National Institute of Standards and Technology), or built by the government and managed by universities (e.g., Oak Ridge National Laboratory by the University of Tennessee).

Today science and technology are part of the global economy. New discoveries and developments spread quickly around the world and are then available universally for applications. It is reasonable for countries that make use of that new knowledge to share in its costs, and especially for big science facilities to be based on international collaborations. There are already such collaborations as the International Space Station (run by the United States and Russia) and the European Spallation Source, a new neutron source to be run by seventeen European countries.

Controversial Science

If science is supported by public funds, then there will be members of the public and of the Congress who will object to the pursuit of particular areas of research, or protest the meaning of the results of the research. Charles Darwin's theory of evolution, published in 1859, still incites ire in those who see a conflict between this theory and their religious beliefs. The use of human stem cells in research has been controversial, again because of religious beliefs. There has been rampant denial of the data showing that the temperature of the atmosphere of the earth is increasing (see figure 3.1), mostly because of the economic price of reducing the production of carbon dioxide from fossil fuels. Lastly, there are those who categorically object to the use of animals in research (chapter 5).

What is to be done when avenues and results of scientific research are controversial? This is an ongoing problem to which there is no simple solution. It is related to the issue of scientific literacy (see next section) in that, if the public understands more about science and technology, then such discussions can be more rational, but not necessarily less contentious.

Science and the Citizen: Scientific Literacy

Science is too important to be left to scientists and ignored by citizens. Too much is at stake: billions of dollars in public funds, the future of the planet, the health of our children and grandchildren. Along with the balance of applied and basic science and the balance of big and little science, there is also a balance of the decision making about public support of science among scientists, citizens, and citizens' representatives.

To take part in the conversation, citizens and their representatives need to have some notion of the nature of science and some knowledge of its most up-to-date status. This education begins with small children, who are naturally curious about the world around them, who peer fervently at a leaf or a bug or a rock, who thrill at mixing vinegar with baking soda, who love dinosaurs, who wonder at the moon. Ideally this curiosity would be nurtured throughout all their lives. An education in basic sciences can be a road to intellectual integrity, to appreciation of the beauty of the natural world, to contemplation of the place of human beings and human values in that world.[20]

There are two fundamental questions about scientific literacy:

1. What should be the curriculum for science and mathematics in schools, from kindergarten through college, and how should that curriculum be taught?
2. What is the means of assessing what American citizens know from that curriculum and from their later experiences and exposures to science? Even among scientists, specialists in one area may know little outside their areas of expertise.

The first question is a recurring effort in American education and, while of great interest, is beyond the scope of this book. However, many of the scientists using this book will teach courses to undergraduates who are not science majors. Those students can become informed citizens and may even become teachers themselves, at various levels of the educational system. These general education science courses are key opportunities to improve science literacy, including the issues of ethics in science.

The second question is also important and complex. In 2014 the Pew Research Center surveyed public knowledge of science topics in the United States. The questions included the following, where the percentage is the proportion of respondents answering correctly:

Uranium is needed to make nuclear energy/weapons (86%).
Distinguish definition of astrology from astronomy (73%).
A light-year is a measure of distance (72%).
Can interpret a scatterplot chart (graph) (63%).
Identify how light passes through a magnifying glass (46%).
Water boils at a lower temperature at high altitudes (34%).[21]

From the considerations in chapter 3, such survey data can depend on the sampling methods, the phrasing of the questions, and the number of respondents. For example, the Pew survey was mostly done online, which limited the sample to those with Internet access, and therefore the sample is necessarily biased in terms of age, educational level, and economic status.[22]

The National Science Foundation regularly assesses the state of scientific literacy and attitudes toward science in the United States, using data based on in-person interviews, which can be expected to be less biased than online surveys. Recent results are similar to those found in the Pew study: in 2014, Americans could correctly answer 4.8 out of 9 science questions. For physical sciences, the true/false questions used were:

The center of the Earth is very hot.
All radioactivity is man-made.
Lasers work by focusing sound waves.
Electrons are smaller than atoms.
The continents have been moving their location for millions of years and will continue to move.[23]

More importantly, are these questions the appropriate ones for judging scientific literacy? Other questions come to mind: What is a molecule? What is DNA? What is the nature of our solar system?

Summary: Science and Public Policy

Science matters to the public. It matters because much of science is supported by public money. It matters because science and the engineering affect the lives of the public in myriad ways: human health, environmental protection, means for warfare and peacekeeping, consumer products, technology and communication, and much more. Citizens and their representatives need to understand the fundamentals of science in order to make decisions about public resource allocations.

Guided Case Study on Science and Public Policy: Scientific Literacy

You are an atmospheric scientist and you are asked to speak to your son's tenth-grade science class on climate change. You show data such as those given in figure 3.1, and you report your conclusion that the data do show that the temperature of the earth is increasing. You then discuss some of the causes and consequences of that change. The next day you receive a phone call from an irate father of one of your son's classmates, who asserts that climate change is not happening and that you are misleading the class.

1. Name the values that are involved and consider the conflicts among them. In chapter 1, the values of life and truth led to the corollary values of the universe, knowledge, and justice. Which of these values are relevant to the issue of climate change?
2. Would more information be helpful? Can you meet with the parent and find out why he feels this way? What if his livelihood depends on the coal industry? Would it help to present more data, from other authorities? Is there any information that would persuade that parent?
3. List as many solutions as possible. Is the parent willing to discuss the issue with you in person? Is he willing to attend a repeat of your presentation? Would it help to hold a group meeting of parents?
4. Do any of these possible solutions require still more information? Do you need to talk to the science teacher? To the school principal? To your son?
5. Think further about how you would implement these solutions. Do you need to find another scientist to join you in this attempt at educating this group of high school students about global warming?
6. Check on the status of the problem. Has anything changed? Are there objections or support from other parents?

7. Decide on a course of action. What if you just cannot change the mind of this parent?

Discussion Questions on Science and Public Policy

1. Discuss the EPSCoR programs. What problems were they designed to solve? What new problems might they create? What would be the considerations if NSF proposed to eliminate its EPSCoR program? Should a scientist be eligible for other funding in addition to EPSCoR?

2. You are developing a new questionnaire to assess public knowledge of science. What are the ten most important questions to ask? Whom might you ask to help you with this list of questions? Do you need a statistician? Who should pay for such a study? What will you do with the results?

3. What is the responsibility of the working scientist to improve civic scientific literacy? In what ways can a scientist assist in this effort—by writing about science for the nonscientist? Helping with science projects at local schools? Volunteering to teach science courses aimed at nonscientists?

Case Studies on Science and Public Policy

1. You are a professor at a state research university. Your senator has offered to write into the federal budget a special appropriation for your college of $100 million for a new program in global warming studies (or whatever program you think would be irresistible). She is even willing to filibuster and to hold up the whole U.S. budget process to make this happen for her alma mater. No proposal and no peer review are required. Is this legal? Is this ethical? What should you do? What should your university president do? What about the argument that this money does not take away from other science funding, but is additional money?

2. You are the director of the National Science Foundation. You have to decide how to allocate new money of $3 billion. A group of physicists wants you to spend $2 billion on a superconducting supercollider. A group of biologists wants you to spend $2 billion on research on the environmental and ecological effects of air and water pollution. A group of chemists wants you to spend $2 billion on centers for new supramolecular nanostructures that could lead to new computer technology.

    a.   How do you proceed to collect information and get advice about these issues?

    b.   How do you make your decision? Should you just divide the funds three ways?

Inquiry Questions on Science and Public Policy

1. The budget of the U.S. government is an indication of what we value as a society.

    a.   Use the Internet or other resources (or phone and ask!) to determine the approximate current annual budgets for:

       The National Science Foundation

       The National Institutes of Health

       The Department of Energy—Basic Energy Sciences

       The National Endowment for the Humanities

       The National Endowment for the Arts

       The Department of Defense—Defense Advanced Research Projects Agency

       One other federal agency of your choice.

    b.   What is the percentage of the total federal budget for each budget in (a)?

    c.   What do the data in part (a) indicate about our priorities as a society?

2. Read more about the SSC project. Why did the SSC fail to survive while the European Large Hadron Collider succeeded? How might the physicists have done a better job of supporting this project?

3. Has scientific literacy in the United States improved over time? Are the analyses that you find convincing? Why or why not?

4. What can you learn about scientific literacy in other countries as compared to the United States? Do you think the results that you find are statistically valid? Is the situation changing with time?

5. There is a long-held belief that the goal for the level of federal support of science should be 3 percent of the GDP.[24] Learn more about this rule of thumb for science policy. Where did this belief originate? What is the level of funding in other countries by this measure?

Further Reading on Science and Public Policy

Budinger, Thomas F., and Miriam D. Budinger. *Ethics of Emerging Technologies: Scientific Facts and Moral Challenges*. New York: John Wiley, 2006.

Claude, Richard P. *Science in the Service of Human Rights*. Philadelphia: University of Pennsylvania Press, 2002.

Flexner, Abraham. *The Usefulness of Useless Knowledge, with a Companion Essay by Robbert Dijkgraaf*. Princeton: Princeton University Press, 2017.

Galison, Peter, and Bruce Hevly. *Big Science: The Growth of Large-Scale Research*. Stanford, CA: Stanford University Press, 1992.

Greenberg, Daniel S. *Science, Money, and Politics: Political Triumph and Ethical Erosion*. Chicago: University of Chicago Press, 2001.

Hiltzik, Michael. *Big Science: Ernest Lawrence and the Invention that Launched the Military-Industrial Complex*. New York: Simon and Schuster, 2015.

Hoddeson, Lillian, Adrienne W. Kolb, and Catherine Westfall. *Fermilab: Physics, the Frontier, and Megascience*. Chicago: University of Chicago Press, 2009.

Holton, Gerald. *Science and Anti-Science*. Cambridge, MA: Harvard University Press, 1995.

Lowen, Rebecca. *Creating the Cold War University: The Transformation of Stanford*. Berkeley: University of California Press, 1997.

Marburger, John H. III, and Robert P. Crease. *Science Policy Up Close*. Cambridge, MA: Harvard University Press, 2015.

McGinn, Robert E. *Science, Technology, and Society*. New York: Prentice-Hall, 1991.

National Science Board. *Science and Engineering Indicators 2016*. Washington, DC: National Science Foundation, 2016.

Riordan, Michael, Lillian Hoddeson, and Adrienne W. Kolb. *Tunnel Visions: The Rise and Fall of the Superconducting Super Collider*. Chicago: University of Chicago Press, 2015.

Stokes, Donald E. *Pasteur's Quadrant: Basic Science and Technological Innovation*. Washington, DC: Brookings Institution Press, 1997.

Weinberg, Alvin M. *Reflections on Big Science*. Cambridge, MA: MIT Press, 1967.

Science and Weapons

To have security against atomic bombs and against the other biological weapons, we have to prevent war, for if we cannot prevent war, every nation will use every means that is at their disposal.

—Albert Einstein, address to the symposium "The Social Task of the Scientist in the Atomic Era," Institute for Advanced Study, Princeton, New Jersey, 1946

War has been in the world for millennia and will be for the foreseeable future. Today's discussions of the ethics of war avoid the moral judgments *good* and *evil* and substitute *just* and *unjust*.[25] The notion of justice introduces a broader perspective, with a balance among competing issues and an acknowledgment that sometimes war may be the only choice. The first ethical issue for a scientist is whether a given war is just. If the war is not viewed as just, then the issue for a scientist is how to avoid participating in that war. If the war is seen as just, then the question is to what extent a scientist is willing to be involved in work that directly supports a just war. However, much work on weapons is general research, not associated with a particular war, and this poses a more general ethical issue.

Scientists value human life. War means killing human beings. Is it ethical for a scientist or engineer to use his or her education, experience, and talents to develop methods for killing other human beings? Of course, almost any new scientific and engineering development can be used to enhance life or destroy life, and it is not possible to know at the beginning whether there will be destructive potential in a particular scientific exploration. Nonetheless, many scientists have faced and will yet face the explicit decision of whether to be involved in weapons research.[26]

Weapons Research in the Twentieth Century

Technology has always been channeled into weaponry, from spears, shields, and slingshots to atomic bombs and computerized drones. Science has been harnessed for weapons or even developed specifically for weapons. The discussion here will begin with the two world wars of the twentieth century and the involvement of scientists in those wars.

When World War I started in 1914, Germany was the epicenter for developments in chemistry and in physics. Great chemists such as Emil Fischer (organic chemistry), Walther Nernst (physical chemistry), and Hermann

Staudinger (polymer chemistry) drew students from around the world. Physicists such as Max Planck (quantum mechanics) and Albert Einstein (relativity) were redefining their science.

In 1909 chemist Fritz Haber, in partnership with Carl Bosch at the German chemical company BASF, had developed a chemical process to convert (*fix*) atmospheric nitrogen gas into ammonia, which then could be converted into the fertilizer ammonium sulfate.[27] The Haber-Bosch process for the synthesis of ammonia was a profound contribution to human survival: it eased food production around the world.[28] The process also contributed to German economic success and made Haber and Bosch into rich and powerful men. When World War I began, the two were quickly drawn into the German military effort because ammonia was needed to make the explosives nitroglycerine and trinitrotoluene (TNT). Then Haber realized that his chemical expertise could be used for even more direct killing by making poisonous gases. The first poison gas used was chlorine, in 1915 near Ypres, France. Haber's involvement in poison gases probably caused the suicide of his wife, Clara Immerwahr, shortly after the Ypres battle. The development of other poison gases followed quickly, including phosgene ($COCl_2$) and mustard gas (($Cl-CH_2CH_2)_2S$), and the pesticide (hydrogen cyanide, HCN, later trade-name *Zyklon B*) that was at first used to fumigate military barracks. Haber's colleague Hermann Staudinger was appalled by the poison gases, and warned the Allies that Germany was going to employ mustard gas. The Allied forces at first condemned the use of chlorine, but soon they, too, used poison gases.

After the war, in 1918, Fritz Haber was awarded the Nobel Prize in Chemistry for his work on the ammonia synthesis. Several other awardees, including Harvard chemist Theodore Richards, who was to receive the 1914 Nobel Prize in Chemistry at the same ceremony, refused to attend because they considered Haber's war work to be immoral.[29]

In the 1930s, Adolf Hitler came to power in Germany. Haber's family was of Jewish origin. In 1933, facing Nazi persecution of his staff and expecting the same treatment for himself, he resigned his post as director and professor at the Kaiser Wilhelm Institute. He found himself abandoned by the country he had served, homeless, financially imperiled, and weak with heart disease. He died in Switzerland in 1934 at the age of 66. The Nazis later used Zyklon B gas in concentration camps to murder the Jews of Europe, including many of Haber's relatives.

Figure 6.5
Chemist Fritz Haber (1868–1934) in about 1905. Courtesy of Archiv der Max-Planck-Gesellschaft, Berlin-Dahlem.

In 1938, physicists Lise Meitner and Otto Frisch and chemists Fritz Stras-smann and Otto Hahn discovered nuclear fission.[30] By 1939, both German and Allied scientists were working to build the first atomic bomb based on fission. Scientists who worked on the bomb included Werner Heisenberg and Walther Gerlach in Germany, and Enrico Fermi, Neils Bohr, J. Robert Oppenheimer, Edward Teller, and Richard Feynman in the United States. Some scientists on both sides refused to work on the bomb, including Lise Meitner and Max Born. Albert Einstein was marginally involved in the American project to develop the atomic bomb, but later regretted even

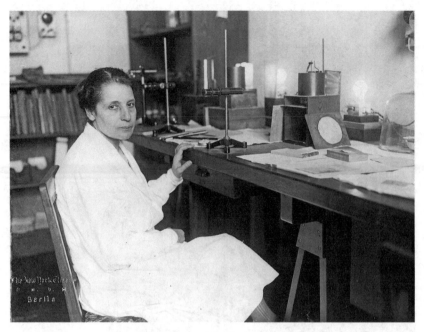

Figure 6.6
Physicist Lise Meitner (1878–1968) in Berlin in about 1931. Courtesy of Archiv der
Max-Planck-Gesellschaft, Berlin-Dahlem.

that involvement.[31] Allied scientists worked on the bomb because they
feared that Hitler would build a bomb first and would use it to conquer the
world.

Werner Heisenberg was a major figure in the development of quantum
mechanics.[32] During World War II, he directed the German project to build
a bomb based on nuclear fission. That project failed.[33] Michael Frayn has
written a play centered on the meeting in 1941 between Neils Bohr and
Werner Heisenberg in Copenhagen, and about the moral dilemma faced
by Heisenberg: whether or not to support Hitler.[34] After the war, Heisenberg and his colleagues promulgated a story that their failure to make a
bomb had been deliberate, that they had thus opposed the Nazis, and that
therefore they were morally superior to the scientists in the United States
and England who had worked on the bomb. However, the evidence is now
clear that Heisenberg wanted Germany to win the war, that he worked
with the Nazis toward that end, and that the German bomb project failed

Figure 6.7
Physicists Werner Heisenberg (1901–1976), Max von Laue (1879–1960), and Otto Hahn (1879–1968) in 1946. Courtesy of Archiv der Max-Planck-Gesellschaft, Berlin-Dahlem.

because of his lack of understanding of the engineering aspects of its construction.[35]

The argument used for poison gases in World War I, and then used again for nuclear weapons in World War II, was that the new weapon would end the war sooner and thus save lives: a utilitarian argument (see chapter 1). It is more likely that poison gases delayed the defeat of Germany in World War I, and thus cost still more lives.[36] Whether the use of the atomic bombs on Hiroshima and Nagasaki saved lives or cost lives has been debated since 1945, with no clear resolution. A newer argument in support of the development of nuclear weapons has been that the level of devastation would be so high that no country will ever use a nuclear weapon again, and so far that has been true.[37]

During World War II, scientists in the United States and in other countries were involved in much research related to the war other than that on the atomic bomb. This included work on radar and cryptography, and further research on poison gases.[38] As just one example, chemist Linus Pauling (see chapter 3) worked on many war-related projects, including a meter to measure oxygen levels in submarines and means to improve explosive powders.[39] The wartime projects in the United States were coordinated and funded by the U.S. Office of Scientific Research and Development (see the Science and Public Policy section at the beginning of this chapter).

After World War II, American scientists remained involved in weapons research. A group called *JASON*, consisting mostly of physicists, was formed to advise the government on military matters involving science and JASON still continues in that role.[40] JASON has worked on problems such as missile defense, and has included Nobel Prize winners Charles H. Townes (laser physics), Murray Gell-Mann (particle theory), and Steven Weinberg (particle theory). Today scientists work in and with government military agencies such as the Defense Advanced Research Projects Agency (DARPA) and the Los Alamos National Laboratory, and for nonprofit federally funded centers such as the Institute for Defense Analyses (IDA), the RAND Corporation, and the MITRE Corporation.

DARPA, founded in 1958, is viewed as "the most powerful and most productive military science agency in the world."[41] The director of DARPA reports directly to the office of the U.S. Secretary of Defense. The DARPA budget of about $3 billion is nearly half that of the NSF and about one-tenth that of the NIH. DARPA issues grants and contracts across all scientific sectors: academic, government, for-profit, and nonprofit. Research supported by DARPA has been instrumental in both military applications (such as drones and stealth aircraft) and nonmilitary applications (including the Internet and global positioning systems).

The U.S. Department of Defense also supports basic scientific research that may have long-term military benefits through the Office of Naval Research, the Army Research Office, the Air Force Office of Scientific Research, and other agencies. The Department of Energy supports the Lawrence Livermore National Laboratory to focus on nuclear weapons. Some university laboratories, such as the Johns Hopkins University Applied Physics Laboratory, conduct the majority of their work on military problems.

Current Issues in the Ethics of Weapons Research

1. Is it more ethical to work on basic research on weapons than on applied research on weapons? An example of applied research on weapons is the development of radar during World War II. However, behind that applied research lay many years of basic research in electromagnetic theory and electronics system development that was and is applicable to a vast number of other peacetime uses, such as radios and computers. Radar itself has many peacetime uses, even though it was developed initially as a weapon of war.

The simplistic dichotomy between basic and applied research was discussed in the Science and Public Policy section and that same analysis applies to weapons research. *Applied* research and *basic* research are not two independent categories. Almost any scientific research can result in military applications. Thus it is not necessarily useful to make a distinction between basic and applied research on weapons.

2. Is it more ethical to work on defensive weapons than on offensive weapons? Where there is an offensive weapon, there is usually a defense against that weapon. Scientists were involved in developing many defensive weapons during World War I and World War II, such as gas masks and radar. Mathematicians in World War II worked to decipher coded messages in order to defend against attacks. Engineers and mathematicians today devise systems and software to defend against ballistic missiles.

Nonetheless, this distinction is a blurred one. For example, even though radar was developed to detect bomber raids against Great Britain, a purely defensive use, today offensive weapons such as cruise missiles use radar for guidance. A scientist may view work on a defensive weapon as being more ethical than work on an offensive weapon, but the distinction is nebulous.

3. Is it more unethical to work on weapons that cause a great deal of collateral damage? Chemical, nuclear, and biological weapons are all *weapons of mass destruction* that generally are seen as more immoral than conventional weapons because their effects cannot be confined to military targets. It has been said that World War I was the chemists' war, World War II was the physicists' war, and World War III will be the biologists' war, because chemicals were used in the first world war, nuclear weapons were used in the second world war, and biological weapons may be key if another world war happens.[42]

Chemical weapons had been banned by the Hague Convention of 1899, which prohibited "the use of projectiles with the sole object to spread asphyxiating poisonous gases," but that did not stop the Germans from using gas warfare in World War I. The decision of Nazi Germany not to use poisonous gas again in World War II may have resulted from deterrence: fear of Allied retaliation with the same.[43] Chemical weapons were banned again in 1993 by the United Nations Chemical Weapons Convention.[44] Nuclear weapons still are not banned by international agreement, but there are treaties to ban nuclear tests and to reduce the size of nuclear stockpiles. Biological weapons—bacteria, fungi, and viruses that carry disease—were banned by the United Nations Biological Weapons Convention in 1972. However, it must be assumed that biological weapons still exist and are still being developed, including genetically modified pathogens. Research continues in most advanced countries on warning sensors for, defenses against, and antidotes to biological weapons.

Conventional weapons, too, can cause considerable collateral damage. During World War II, conventional weapons used in the German bombing of England killed about 100,000 civilians, and the Allied bombing in Germany killed about 780,000 civilians.[45] Scientists and engineers are now engaged in research to increase the precision of bomb and missile delivery in order to reduce damage to nonmilitary targets. With higher precision, smaller warheads can be used, further decreasing collateral damage. However, injury to noncombatants can occur at times other than during an attack. For example, a century after World War I, buried bombs and gas canisters continue to surface all over Europe, and continue to injure and kill innocent people. Fifty years after the war in Vietnam and Cambodia, land mines still explode and cause harm.

The advent of *cyber warfare* in the twenty-first century has introduced a new kind of weapon of mass destruction. Cyber weapons are computer codes, not explosives. They can serve defensive and offensive purposes, and they can be incredibly destructive. One of the more famous cyber incidents was the introduction in 2010 of the Stuxnet computer virus into the Iranian computer network controlling centrifuges that separate isotopes of radioactive materials.[46] The Stuxnet virus is thought to have destroyed at least 20 percent of the centrifuges, setting back the Iranian nuclear program, but with no harm to humans or other equipment. There was no collateral damage in this instance, but computer attacks could cause significant injury to noncombatants because computers control many critical activities in advanced countries. River dams can be opened remotely, bank accounts can be emptied, air traffic control communications can be shut down, subway trains can be made to collide, traffic lights can be randomized, and hospitals can be darkened—all without exploding a single traditional weapon. Cyber warfare can cause mass destruction.

4. Is it more unethical to work on weapons that cause a prolonged and painful death? Quick death in war may seem more ethical than slow and painful death. A bullet or a grenade usually kills quickly, while mustard gas or nuclear radiation can cause agony before death. On the other hand, bullets and grenades can result in lingering death and severe disablement and disfigurement. Many British soldiers returning from World War I "no longer had noses, eyes, jawbones, cheekbones, chins, ears, or much of a face at all."[47] Surgeons developed techniques to restore faces, and, when those techniques failed, sculptors made tin face masks for the disfigured soldiers to wear.

The two criteria of avoiding collateral damage and avoiding painful deaths are not easy to apply, and there is no easy classification of weapons into *ethical* and *not ethical*. Scientists thinking about weapons research must live with this ambiguity.

Summary: Science and Weapons

Scientists in the United States can reasonably hope not to face the ordeal of being in the grasp of an evil government that forces a collaboration in repugnant acts of war. However, scientists may find themselves working for companies or research groups that do weapons research: basic or applied, offensive or defensive, limited to combatants or not, humane death or not, consequences known or not known. Decisions about working on weapons can be made thoughtfully, using the skills developed in chapter 1.

Guided Case Study on Science and Weapons: Chemical Weapons

You are an American industrial chemist. Your company has a new, secret contract with the Department of Defense to develop a chemical weapon that defeats all known gas masks. Do you agree to work on the team to synthesize this weapon?

1.   Name the values that are involved and consider the conflicts among them. Human life was the first value listed in chapter 1. Whose lives are in danger? American lives? Other lives?

2. Would more information be helpful? Is there an immediate use for the weapon, or is it meant for future availability? What is the scope of the weapon: will it be used in restricted areas, or will it be a weapon of mass destruction? Does the development of this weapon violate any international agreements?

3. List as many solutions as possible. You can work on the team or not work on the team. If you work on the team, can you set some conditions to your involvement? For example, can you ask to work on defenses against the new weapon rather than on the weapon? Can you argue that the weapon should be designed to cause temporary disablement rather than death?

4. Do any of these possible solutions require still more information? If you decline to work on the team, what is the impact on your career? Will you lose your job? Will you lose opportunities for advancement?

5. Think further about how you would implement these solutions. Do you need to talk to your colleagues? To your supervisor? To your family? Can you read accounts by other people who have faced such a problem?

6. Check on the status of the problem. Has anything changed? Perhaps you have found an opportunity for a job elsewhere. Perhaps the weapon will not be a lethal one.

7. Decide on a course of action. Other scientists of integrity have worked and still work on weapons. It is not necessarily an unethical path.

Discussion Questions on Science and Weapons

1. Consider the use of computer weapons against the computer networks of an adversary. Against what types of targets and to what ends should offensive cyber weapons be employed? What are some issues to consider for cyber defenses?

2. Improvements in computer engineering have made it possible to construct drone devices that are controlled from afar and that can collect information and deliver weapons to targets. Are the ethical issues with the use of drones different from those with human soldiers?

3. Over several decades, research was conducted on nuclear weapons that minimized blast and maximized the highly penetrating neutron flux, with the intention of killing living beings while leaving buildings and military equipment largely intact. The attacking soldiers could safely

occupy buildings within a few days, after the radiation effects were gone. These were called *enhanced radiation weapons* or *neutron bombs*. None exist now in the American inventory. Discuss the ethical implications of such weapons. Under what conditions, if any, might such weapons be used? Does the use of the neutron bomb differ ethically from the use of a poison gas that causes a painless death? What about collateral damage?

Case Studies on Science and Weapons

1. The first and only time that a nuclear weapon has been used in a war was when the United States dropped atomic bombs on the Japanese cities of Hiroshima and Nagasaki in 1945. President Harry S. Truman's decision to approve the use of nuclear bombs has been debated ever since. Outline reasons why scientists should have worked with the American government to develop the atomic bomb. Then outline reasons why scientists should *not* have worked with the American government to develop the atomic bomb. Discuss why the bombs should or should not have been used against Japan. Discuss whether the balance of nuclear power in the world may have (so far) prevented a third world war.

2. Imagine that you are Werner Heisenberg in 1941. You love physics, you love your job as Herr Professor, and you love Germany. The Nazis are sending your university colleagues to concentration camps: What do you do? You are required to salute and say "Heil Hitler" before every lecture: Do you comply? You visit your old friend Neils Bohr in Copenhagen, now occupied by Nazi forces who have planted the swastika flag everywhere: What do you say to Bohr? Many of your fellow physicists have left Germany: Why do you stay? After the war, how do you explain why you collaborated with the Nazis?

3. In 2015, the American Psychological Association (APA) admitted that several of its high-ranking officers (including the ethics officer) colluded with the U.S. military and the Central Intelligence Agency in providing justification for the torture (such as waterboarding and sleep deprivation) of prisoners of war at Guantanamo Bay, Cuba, and in Iraq and Afghanistan.[48] The motivation of these APA officers was to support the many psychologists who are directly employed by the U.S. military. Compare the involvement of these social scientists in

military activities with the involvement of chemists and physicists in military activities.

4. You are a chemist working on a new compound that has uses for disease treatment and also for poison weapons. The Department of Defense contacts you and offers support for your research. The National Institutes of Health also offers you support. Does it matter to the direction of your research who provides the money? What if the money from the Defense Department is ten times as much as that from NIH? What if you accept both offers: How do you manage the differing objectives? How does your situation compare to that of Fritz Haber in 1914?

Inquiry Questions on Science and Weapons

1. The chemical defoliant Agent Orange was used in the 1970s during the war in Vietnam. What is the molecular structure of this compound? How was this chemical discovered and developed? Which chemical company was involved? What were the ethical issues in its use? Is a defoliant a *chemical weapon*? What were the unintended consequences?

2. Read the play *Faust* by Johann Wolfgang von Goethe. How can a scientist be seduced into a *Faustian bargain*? Are there examples of such in the cases discussed in this chapter?

3. Scientists can be charged with war crimes. Find out what happened to German physicists Philipp Lenard and Johannes Stark, both avid Nazis, after World War II. How did Werner Heisenberg avoid being charged with war crimes? If the Germans had won the war, could the atomic physicists who worked on the Manhattan Project have been charged with war crimes? Is a scientist's work on weapons criminal only if his or her side loses the war?

4. During work on the atomic bomb in about 1941, physicists realized that a nuclear weapon could be devised that was vastly more powerful than the atomic bomb, that uses nuclear fission to cause a nuclear fusion explosion. The physicists were then divided between those who thought that this *H-bomb* should be developed, and those who thought that it should not be developed. Learn more about this debate and its outcome. Which physicists were on each side? Who won the debate and how did they win?

5. Wernher von Braun was a German rocket scientist who developed the V-2 rocket that was used against Great Britain. He was a member of the

Nazi party. After the war, he surrendered to the American forces. Learn more about Wernher von Braun. What happened to him after that? What are the ethical issues that arose in his life? What ethical issues arose for the American government after his surrender?

Further Reading on Science and Weapons

See also biographies of scientists in appendix B.

Bernstein, Jeremy, and David Cassidy. *Hitler's Uranium Club: The Secret Recordings at Farm Hall.* 2nd ed. Göttingen: Copernicus, 2001.

Borkin, Joseph. *The Crime and Punishment of I. G. Farben.* New York: Free Press, 1978.

Bridger, Sarah. *Scientists at War: The Ethics of Cold War Weapons Research.* Cambridge, MA: Harvard University Press, 2015.

Cornwell, John. *Hitler's Scientists: Science, War, and the Devil's Pact.* New York: Viking, 2003.

Jacobsen, Annie. *The Pentagon's Brain: An Uncensored History of DARPA, America's Top-Secret Military Research.* New York: Little, Brown, 2015.

Schweber, Sylvan S. *In the Shadow of the Bomb: Bethe, Oppenheimer, and the Moral Responsibility of the Scientist.* Princeton, NJ: Princeton University Press, 2000.

Tenes, Peter S. *The Just War: An American Reflection on the Morality of War in Our Times.* Chicago: Ivan R. Dee, 2003.

Conflicts of Interest

No servant can serve two masters: for either he will hate the one, and love the other; or else he will hold to the one, and despise the other.
—Luke 16:13, The Holy Bible, King James Version

A conflict of interest can exist when a person has two or more obligations or responsibilities or interests, and these interests are not completely independent of one another. The fulfillment of one interest affects the fulfillment of the other interest, so that it is hard to satisfy both at the same time and with integrity. The resolution of a conflict of interest is an ethical issue, to be approached with the problem-solving methods developed in chapter 1.

Conflicts of Interest for Scientists

In doing science, a scientist seeks to find the truth, as discussed in chapter 3. Other interests can come into conflict with the search for truth: need for money for the research, desire for professional advancement or for fame, loyalty to family and friends and to students and collaborators, allegiance to a college or university, devotion to country. Profound conflicts of interest can arise when a scientist is asked to work on the development of weapons, as discussed in the section on Science and Weapons in this chapter: conflicts among devotion to science, love of country, and care for humankind.

Conflicts of interest regularly arise in a scientist's tasks of reviewing manuscripts and proposals (see chapter 4). For example, a scientist who is reviewing a proposal for funding will have a conflict of interest if that proposal was written by a former student. Hence, proposals for National Science Foundation funding must include lists of recent collaborators and former students, and those people are not asked to review that proposal. In addition, journals and granting agencies will allow the author of a paper or proposal to list persons who he or she does *not* want to serve as reviewers and will honor such requests.

Another example is conflict between one's professional responsibilities and the interest in one's own financial status. Suppose that you are to choose a new general physics textbook that will be used by a thousand students every year at your university, and the representative of one textbook offers you a consulting fee if you select his textbook. Your responsibility to choose a good book is then in conflict with the opportunity to make money for yourself. Sometimes the conflict involves not money, but a personal relationship. Suppose that you are to choose a new physics textbook, and your best friend has published just such a book.

The conflict could also be one of time: a *conflict of commitment*. Suppose that you are a full-time computer science professor at a university, and you start your own consulting business on Internet security. Your new business requires attention that you would otherwise spend on your teaching, service, and research at the university. Conflicts of time and commitment are then conflicts of interest if one commitment interferes with the time and energy that you have available for the other commitment.

Financial conflicts of interest are easier to ascertain and to control than are other conflicts, and often are regulated by laws that explicitly prohibit some activities and yet permit other activities so long as they are properly disclosed. The federal government has established policies for its grant-giving agencies (NSF, NIH, FDA) that require that the institutions through which the grants are managed (colleges, universities) each develop a process to review and manage any financial conflicts of interest.[49] The threshold at which a conflict is defined is $10,000 per year of financial interest that is related to the project, but how that $10,000 is calculated can get complicated, and rules vary among agencies. For example, an organic chemist working on synthesizing a new drug to combat HIV may have a conflict of interest if he or she owns stock in the company that provides the leading current HIV drug. The resolution of a financial conflict of interest can vary from withdrawal of the researcher from the project, to withdrawal of the researcher from the financial interests that caused the conflict, to oversight of the researcher by an independent party. If there are human subjects involved in the research, then the researcher's financial interest in the outcome of the research must be included on the informed consent form (chapter 5, section on Research with Human Participants).

Federal employees, including scientists, are under strict conflict of interest laws in the federal criminal code, violations of which are felonies.[50] Federal employees are allowed to accept only token gifts in the course of their duties or from job-related activities, with "token" defined as worth $20 or less per occasion and not more than $50 per year. However, to keep it simple, most government agencies adopt a policy of no gifts whatsoever. NSF employees are forbidden from even accepting a free lunch from potential grant seekers. Attendance of federal employees at events (e.g., conferences or trade shows) has to be paid for by the agencies or, if free, approved ahead of time by the agencies. Scientists who serve as paid members of review panels at NSF or NIH, or who serve on government advisory panels, are considered to be special government employees and are subject to the federal code. For example, a scientist could not receive a speaking fee for a talk on "My Life on an NIH Review Panel." Naturally, federal employees are prohibited from lobbying activities, since they cannot lobby the government by which they are employed.

Conflicts of Interest for Institutions

Figure 6.2 shows the dramatic increase in corporate (industrial) support of research, much of which is external to the companies and is done through contracts with universities. The degree of collaboration between companies and universities has increased since the 1980 Bayh-Dole Act that gave academic researchers the right to license to industries their discoveries made with federal research funds.[51] The Federal Technology Transfer Act of 1986 made it possible for researchers working in the U.S. government to license their results to industries.

Working together, academia and industry can advance science and technology and contribute to the growth of the economy.[52] However, their different missions can lead to conflicts. Colleges and universities have as their mission the transmission of knowledge by teaching and the generation of knowledge by research. Usually they are *nonprofit* and are not in the business of making money, whereas companies exist to make money. Professors and their students expect to share their research results by publishing openly in journals and speaking freely at conferences. Companies expect to retain full rights to the products of research that they have supported, and may want those results to remain confidential. University researchers who work with companies or who start their own companies will encounter conflicts between the culture of open research and the needs of companies to preserve their prerogatives.

Companies also may expect that the research that they support will reflect favorably on their products. In 1953, the tobacco industry set up the Tobacco Industry Research Committee (later called the Council for Tobacco Research) in order to promote research that supported the use of tobacco.[53] The American Natural Gas Association has attempted to control information on the dangers of fracking.[54] A scientist who accepts research funds from companies may feel pressure to reach a particular conclusion, or may find that research conclusions are suspect because of the source of funding, or may even find that the results are suppressed by the company. One good practice for scientists is to always disclose the sources of funding for their work in their publications, whether or not the journals require them to do so.[55]

Guidelines for Managing Conflicts of Interest

Conflicts of interest are inevitable and unavoidable, but their existence does not mean that unethical conduct is unavoidable. *The general rule for managing conflicts of interest is that they be acknowledged immediately, honestly, and openly.*

Transparency in itself may solve many of such problems. Once the conflict is known, every effort can be made to separate the interests. Sometimes the solution is straightforward: you can withdraw from the situation. For example, you simply decline to review a paper written by your spouse.

Good institutional policies can mitigate the conflicts. Conflicts can be made known ahead of time so that problems are avoided. For example, university requirements that outside consulting be limited to one day per week and be formally reported annually make it easier to manage conflicts of commitment. Similarly, when you disclose the source of research funding, you may not dissolve a conflict of interest completely, but you do make the possible effects of the conflict easier to analyze.

Summary: Conflicts of Interest

Conflicts of interest arise for a scientist in doing research, in dealing with other people, in working within institutions, companies, and organizations. Open discussion of the nature of conflicts of interest, especially among mentors and mentees, can raise awareness and prepare minds. Institutional policies for reporting and addressing conflicts of interest are essential.

Guided Case Study on Conflicts of Interest: Textbook Author

Professor Goodbook has written a textbook in general chemistry. He teaches the general chemistry course and can require that his 250 students buy his textbook. For every $200 textbook that is sold, Goodbook makes $20. Are there ethical issues here? What are possible solutions?

1. Name the values that are involved and consider the conflicts among them. Think about how truth and justice enter into this problem. Can you articulate the conflict of interest that Goodbook faces?
2. Would more information be helpful? Is Goodbook's textbook the only such book on the market? If not, are there reviews of such textbooks that support claiming that Goodbook's textbook is the best book on the market? Does the conflict of interest change if his book is the only book or the best book available?
3. List as many solutions as possible. It can be argued that Goodbook will teach best from his own book and that students will then learn best, even if there are other books available, and thus that he should select his own book. That still leaves the appearance of a self-serving decision. Could Goodbook assign all the royalties from book sales to his

own class to the university for which he works, so that he is not profiting from the textbook decision? Are there other solutions?

4. Do any of these possible solutions require still more information? Goodbook will need to track the books sold to his class and find a means of assigning those royalties to the university, so he may need to talk to his publisher.

5. Think further about how you would implement these solutions. Should Goodbook make a statement on the syllabus about the royalty assignment, in order to avoid problems with those who wonder about his textbook choice? Should he talk with his department chair?

6. Check on the status of the problem. Has anything changed? When he contemplates publishing a new edition of his book, should Goodbook require that students buy that new edition? Are the issues the same or different?

7. Decide on a course of action. Is the problem solved?

Discussion Questions on Conflicts of Interest

1. A scientist could have an intellectual conflict of interest if the results of a new study by that scientist are in conflict with the results of a previous study by that same scientist. What are the temptations and their resolutions?

2. What conflicts of interest arise when undergraduate or graduate students at a university get involved in research with a professor and that research is supported by a company?

Case Studies on Conflicts of Interest

1. In March 2008, the governor of Maryland was on the University of Maryland, College Park, campus to celebrate the establishment of a company called *Zymetis*. Zymetis was founded to exploit the discovery by a Maryland professor of a means of converting biomass into ethanol. This company had its laboratories on the university campus. What ethical issues might arise from locating a company on a university campus? Would it matter whether the university was public or private? How could you prevent materials and equipment that were bought with non-Zymetis money from being used by the company, or vice versa? How about the use of student labor by the company? Does the professor who is CEO of Zymetis have conflicts of interest?

2. Professor Cooperstein teaches one section of a large introductory chemistry class, for which a new textbook will be selected by the team that teaches the four sections of the course. The textbook salespeople are very interested in this decision, since the total sales to this class are about three thousand books per year. Two textbooks are being considered and the representatives of both publishers want to make presentations to Cooperstein and her colleagues.

    a. Both representatives offer to provide lunch to the teaching team during their presentations. Is that acceptable?

    b. One representative offers four free books from his catalogue to each team member. Is that acceptable?

    c. One representative offers to publish Cooperstein's new monograph. Is that acceptable?

    d. One representative offers free trips to the next American Chemical Society meeting for the team members. Is that acceptable?

    e. Does it matter whether the enticements in parts (b)–(d) are offered with or without requiring that the representative's book be selected?

3. Professor Gibbs is reviewing a proposal for a grant from NSF in his field of statistical physics. He knows that funds in that field have been reduced and that if this proposal is funded, then his own renewal proposal, to be submitted in a month, will have less chance of being funded. Does he have a conflict of interest? If so, what is to be done?

4. Professor Smith at University X and Professor Jones at University Y just do not like each other. This has been the case since their marriage was dissolved ten years ago. Should they be allowed to review one another's papers or proposals?

5. Professor Washington and Professor Lee have been close friends since graduate school. They work in the same area of research and often talk together about their work. They visit one another's homes on vacations and sabbaticals. Should they review one another's papers and proposals? How close must a relationship be before a conflict of interests exists? What if they are the best judges of one another's work?

6. Dr. Maxwell and her husband were both recruited to the Department of Chemistry, she as a full professor and he as an assistant professor. Three

years later, the tenure discussion for the husband is to be discussed in a faculty meeting. Maxwell is sitting in the meeting. What should happen and who is responsible for making it happen?

Inquiry Questions on Conflicts of Interest

1. Find out the rules on conflicts of interest and on conflicts of commitment at your institution. Are there rules about stock ownership and its disclosure? For example, can a researcher who is evaluating chemical treatments for cancer own stock in a company that markets one of the chemical treatments? Who is responsible for monitoring these rules?
2. Has there been a controversy over a conflict of interest in a research project at your institution? How was it resolved?

Further Reading on Conflicts of Interest

Charrow, Robert P. *Law in the Laboratory: A Guide to the Ethics of Federally Funded Science Research*. Chicago: University of Chicago Press, 2010.

Greenberg, Daniel S. *Science for Sale: The Perils, Rewards, and Delusions of Campus Capitalism*. Chicago: University of Chicago Press, 2007.

Resnik, David B. *The Price of Truth: How Money Affects the Norms of Science*. New York: Oxford University Press, 2006.

Intellectual Property

The Congress shall have Power. ... To promote the Progress of Science and useful Arts, by securing for limited Times to Authors and Inventors the exclusive Right to their respective Writings and Discoveries.
—The United States Constitution, Article I, Section 8

The term *intellectual property* indicates that the products of the mind are of real value and are to be treated as property in the same sense as material property. You cannot take and use my car without my permission, and likewise you cannot take and use my ideas, or my words, or my inventions with impunity. Indeed, intellectual property issues are as much legal as ethical: intellectual properties are defined and protected by law in all developed countries. For a scientist, intellectual property is an extremely important matter in protecting the products of one's own work, and in making fair use of the contributions of others.

Intellectual property can be divided into four types: trade secrets, trademarks, copyrights, and patents, in increasing level of legal protection.

Trade Secrets

Trade secrets are proprietary designs, formulas, computer programs, and other such significant creations that have been developed and kept confidential because of their economic advantages to businesses, but have not been patented. The most famous trade secret is the recipe for Coca-Cola, invented in 1886 and never revealed, as it would have to be if it were patented. As long as a trade secret is kept confidential, it is protected by law in that employees and others can be required to maintain the secrecy by nondisclosure and noncompetition agreements.

In addition, illegal theft of trade secrets is covered by the U.S. Uniform Trade Secrets Act of 1979 and 1985, which makes standard the laws on trade secrets within the states, and by the Economic Espionage Act of 1996, which makes it a federal crime to steal trade secrets across state boundaries or in benefit of a foreign power. Such *industrial espionage* could include hacking into the company computer or stealing documents. The Trade Secrets Act requires that federal employees do not reveal trade secrets that they learn in the course of their duties. Trade secrets are protected by laws only so long as they are kept secret, and otherwise have no expiration time period.

However, it is not illegal to reverse engineer in order to obtain a trade secret. For example, you can buy a bottle of Coca-Cola and attempt to determine its composition by chemical analysis. Perhaps you want to use the formula for Coca-Cola to *benchmark* what is standard industry practice as you develop a new beverage.[56] You then legally can use that information to make your beverage, but you cannot call your beverage Coca-Cola because that is a registered trademark (see the next section).

The advantages of a trade secret are that it requires no legal application to secure it, and that it can remain a secret forever, thus keeping the benefit of its use to the owners and licensees in perpetuity. The disadvantages are that the owner must be vigilant to maintain the secrecy, and that anyone who honestly figures out the secret is free to use it.

The methods and practices of a working scientist, even if they were developed by that scientist and are different from methods used by others, are not regarded as trade secrets. Trade secrets could become an issue for scientists who devise new products or processes that have commercial value and start their own companies or who work for or with companies. A chemist might develop a process for processing a polymer. A physicist might devise a microvalve for surgical use. Then the inventing scientist will have to decide whether to keep the procedure as a trade secret, or to seek a patent (discussion follows).

Trademarks

Trademarks are names, symbols, or logos that are developed to designate a particular brand or product. *Service marks* serve the same purpose for services as opposed to products. Trademarks and service marks can be registered to provide legal protection for their owners and to protect consumers from fraud. In science, common trademarks include *Kimwipes* and *Pyrex*. Scientists are unlikely to need to register a trademark unless they start a business—for example, making a new analytical instrument—and then want a trademark for protection of their product.

The superscript symbols ™ and ℠ indicate unregistered trademarks and service marks; the symbol ® (R in a circle) indicates a *registered* trademark or service mark. These symbols are inserted just after the name or logo. Anyone can use the superscript ™, but only registered users can use the symbol ®. Trademarks and service marks can be registered at state and federal levels. Registration in foreign countries must be done in each country. Federal registration is done by the U.S. Patent and Trademark Office (USPTO) and requires a fee of about $400 and a search to be sure that the mark is not already registered. The USPTO website has full information on how to establish a trademark. To maintain the registration, the mark must be used continually; renewal must be made in the fifth and sixth years and then every ten years. In principle, trademark protection can be maintained indefinitely. Once the mark is registered, the owner has protection against the use of the mark by others. Registered trademarks can be sold or licensed. Unregistered marks are not protected from use or theft.

Copyrights

Copyrights provide legal protection for written expositions such as books and articles and for artistic compositions such as music and painting, so that the use of the materials is controlled by the persons who created them. Scientists are mainly concerned with written documents such as research papers and technical books, and with electronic products such as computer programs. Copyrights protect original works of authorship, where *original* means that the author did the work himself or herself, and is not the same as *novel*, which means that no one else has ever done something similar.[57] Novelty is required for patents, but not for copyrights. Originality is a weaker requirement than novelty.

The U.S. Copyright Act of 1976 protects the *expression* of ideas, but does not protect the ideas themselves and thus does not protect scientific discoveries or facts.[58] The copyright precludes another person from copying or paraphrasing the written language, but does not prevent another person from pursuing the same subject, and even allows another person to use the exact written material in certain cases of *fair use*, as will be discussed. This legal protection is weaker than that of a patent, since the patent prevents another person from independently developing the same product. While others may use a copyrighted work legally, the ethical requirement is that users give appropriate credit to the originators, as discussed in chapter 4.

Written material is copyrighted automatically by common law—you do not need to formally register the copyright. However, your legal rights are better protected and you have a better chance of winning a case of infringement against your copyright if you register your material. You can register with the U.S. Government Copyright Office, which is overseen by the Librarian of Congress; foreign countries each will have a similar procedure. The form is online at the website of the Copyright Office and the fee is about $55. Two copies of your copyrighted material must be submitted for donation to the Library of Congress. Since 1978, a copyright is in effect until seventy years after the death of the owner. A copyright is indicated by labeling the material with "Copyright *Year* by *Name of Author or Owner*" or with "© *Year* by *Name of Author or Owner*" ("by" is not always included in the copyright claim). For example, for this book the proper form is "© 2017 Massachusetts Institute of Technology."

Fair use is a term used in the U.S. Copyright Act to mean the legal use of copyrighted materials for purposes such as criticism, review, or teaching. Fair use involves four considerations.[59] First, the use should be nonprofit; for example, for research or teaching. Second, the work used should be factual in nature, as opposed to creative in nature. For example, using dates from

Winston Churchill's *The Second World War* would be allowed (with attribution), but using his same language would not be allowed except as a quote, within quotation marks and with a reference. Third, even for quotes within quotation marks, the fraction of the work used should be small—typically less than seven lines. Fourth, the use of the work should not reduce its commercial value. A scientist can copy a copyrighted journal article for his or her personal scholarly use, but cannot make copies for an entire conference and distribute them. The distribution of copies reduces the income of the copyright holder. These four criteria are vague, interrelated, and subject to different interpretations by the courts. The use of copyrighted material that exceeds the limits of fair use requires the permission of the copyright holder, and probably requires payment to get that permission.

The advent of the Internet and its multitude of websites has created new challenges for copyright protection. The Digital Millennium Copyright Act of 1998 (DMCA) protects copyright holders from unauthorized use of their material on websites by requiring that the material be removed when a DMCA Takedown Notice is filed. In 2013, "Mainstream copyright owners [sent] takedown notices for more than 6.5 million infringing files, on over 30,000 sites, *each month.*"[60] This system is not working, since material regularly reappears shortly after takedown and the procedure has to be repeated. Congress will surely revisit this problem.

For you as a scientist, the copyright law is important first in protecting your own work and then in assuring that you take proper precautions in using the work of others. In copyrighting your own work, there are three cases of interest.

1. You write an article that is published in a journal. Scientific journals require that you and your coauthors assign the copyright to your article to the journal by signing a form. The journal then obtains and owns the copyright. You lose the power to reuse your own work! You cannot use figures from your own article in another article without the journal's permission. You cannot put a copy of the article on an open website without the journal's permission. Journal policies vary in their details, so you will want to refer to that journal website when you want to use material from one of your own articles.

The Printing Law of 1895 declared that the U.S. government and its employees cannot hold copyrights to material produced in the scope of government business: such material belongs to the public. If a federal employee is an author or coauthor of a scientific publication, the copyright for that paper cannot be transferred to the journal. For authors who are not federal employees but whose work is supported by federal funding, the copyright law depends on whether the funding is a grant (for which there are no specified deliverables) or a contract (in which deliverables are specified). Scientists working under federal grants are entitled to hold copyrights to the products of those grants, but will usually transfer those rights to the journals in which they publish. The rights to copyrights of work produced under federal contracts, however, will depend on the terms of the contracts (see item 3). Small science is generally funded by grants, and big science is generally funded by contracts (see earlier Science and Public Policy section).

2. You write a book, computer program, or other work, on your own initiative. If you write the book at home, on your own time and using your own resources, then the copyright is yours alone. If you write the book at work, using work resources, the copyright may not be yours, even if the project was done on your own initiative. In the past at a college or university, the copyright has belonged to the authors, even though the authors may have used college or university resources (computer, library, copier) to produce the book or computer program. Likewise, the rights to course syllabi have previously been kept by faculty members. Now these policies are changing, and colleges and universities are beginning to retain the copyrights at the level of the institution, and may or may not allow the originating faculty member to use the material.

The same may be true of scientists in industry, who are wise to check on particular company policies. Government employees cannot hold copyrights for work done on the job, as discussed previously. However, industrial or government employees can hold copyrights to material that they produce on their own initiative and on their own time. Any use of work facilities (library, copier, computer) will jeopardize that independence.

3. You write a book, computer program, or other work, as specifically directed by your employer or as required under a contract with a company or the government. Such activity is called *work for hire* or an *assigned duty*. When you develop material *in the scope of your employment*, then your employer owns the copyright. A scientist working in industry will not own the copyright for material produced during his or her working day. Government employees cannot hold copyrights for work done on the job. Such a case would be very unusual in a college or university, since academic scientists work very independently of management in both research and teaching and are not likely to be directed in their work, but see preceding item 2.

In your use of the copyrighted work of other people, there are two common cases:

1. You want to use copyrighted work in your teaching. If you want to use copyrighted material in teaching, then you can do so—but only to a certain extent. You are allowed to copy small segments of a book for your students, but only if the segment is sufficiently small and only if it is a spontaneous, one-time event. You cannot copy large parts of the book, and you cannot use the same parts of the book every year, because you are then reducing the compensation to the author of the book.

Now there exist course websites, onto which you can load documents and on which you can provide links to other websites, raising new issues of copyright law. Congress addressed the problem with the Fair Use Guidelines for Educational Multimedia of 1996, and the Technology, Education Copyright Harmonization Act (TEACH Act) of 2002. Material on the Internet is copyrighted, by common law even if it is not registered. Links to that content are not copies and therefore are not copyright infringements. Copyrighted documents may be put on course websites with access limited to students in the class, if the students are instructed not to distribute the documents and if the access is limited to a few weeks. You should contact your institution's intellectual property officer (usually a library staff member) for detailed policies and instructions.

2. You want to quote the copyrighted work of others in your own work. Recall from chapter 4 that you have an ethical obligation to give credit to the ideas of others upon whose work you have built. You can quote from others' work or paraphrase from their work (with proper attribution) and you will not be in violation of their copyrights. However, if you want to use tables, figures, or lengthy written passages taken directly from other works, then you will need permission from the copyright owners. As discussed in item 1, you will even need to get permission to reuse your own published work if the copyrights have been transferred to publishers.

Patents

Patents provide legal protection for new inventions and ideas. While copyrights protect the unique ways of expressing ideas, patents go further and protect the ideas themselves. In exchange for this protection, the inventor must disclose full information about the new design, formula, or device.

In order for an idea to be patentable, it must be a new idea that is not obvious, it must be useful, and that usefulness must be established. The criterion of "not obvious" means that an expert in the field would not immediately come up with the invention. In chapter 3, the importance of keeping detailed laboratory records was explained. Those records are essential when a patent application is prepared, to establish the originality and the date of the discovery.

Patents are issued in the United States by the U.S. Patent and Trademark Office (USPTO), based on the Patent Act of 1952 and its subsequent revisions. The time periods of patent validity have changed over time, but currently a new patent is in effect for twenty years after the date the application is filed (with some exceptions—see USPTO website). Once the patent is issued, no one else may make, use, or sell that invention in the United States for the term of the patent, without the permission of the patent holder. Patents may be bought and sold (*assigned*), or licensed.

Patent laws in other countries differ from those in the United States. First, when there is a dispute about a patent in the United States, the patent is awarded to the person who first *had the idea* and began its development. In other countries, the disputed patent goes to the first person to *file* for a patent for the invention. In the United States, an inventor may file a *provisional* patent application that is not examined by the USPTO but that preserves the rights for foreign patents. Second, when an inventor in the United States publicly discloses the new invention, he or she then has one year from that date to file the patent application. In foreign countries, any public disclosure precludes the application for patent rights. Thus an inventor should be very careful about public disclosures (such as conference presentations), should file a provisional application as soon as possible, and should move quickly to the final application.

Gordon Gould (1920–2005) was an American physicist who figured out how to make a laser and who invented the term *laser* from *Light Amplification by Stimulated Emission of Radiation*. At first, Gould lost out on the American patent for the laser because he waited too late to file an application,

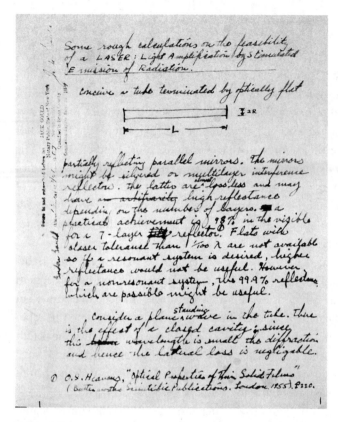

Figure 6.8

A page from the 1957 laboratory notebook of Gordon Gould, describing his idea for the laser. The page is titled, "Some rough calculations on the feasibility of a LASER: Light Amplification by Stimulated Emission of Radiation." Note the notarization in the left margin. With permission of the AIP Emilio Segrè Visual Archives, Hecht Collection.

but he was able to use his dated and notarized laboratory notebooks to get part of the patents—and millions of dollars—after a thirty-year legal battle.[61] He could prove that he had the idea for the laser first, even though he did not file first.

To file a patent application, you should hire an attorney. You could proceed on your own, but this probably is not wise. Your attorney must by law be admitted to practice with the USPTO. If you work for a university, government, or industrial laboratory, the organization (via its Office of Technology Transfer) will have an invention disclosure form to start the process, and then will support you through the rest of the process and provide a lawyer. The patent lawyer will prepare the application with your help, and the USPTO will assign a patent examiner to your application. The process can take years and cost tens of thousands of dollars.

The patent probably will be the property of your employer, but you may have some rights to royalties, depending on the policy of the university or the company. The Bayh-Dole Patent and Trademarks Amendment Act of 1980 allowed patents from research done under federal grants and contracts to be held by the inventors and their universities or companies, and the Federal Technology Transfer Act of 1986 did the same for government employees and laboratories. The exact policies and terms for sharing revenues are worked out within each organization and vary widely, but those policies and terms can often be renegotiated.

There are some discoveries that cannot be patented. You cannot patent laws of nature or naturally occurring organisms or molecules. However, you can patent nonobvious uses of those laws, organisms, or molecules. For example, genetically modified plants and animals can be patented. However, gene sequences have been denied patents because their usefulness was not established.[62]

Ownership of Records and Data

Samples and materials collected during the course of research will contain information and thus are *intellectual property*, but are not included under copyrights or patents. Such materials and laboratory records (notebooks and computer files) are considered the property of the employing organization and are not the property of the individual scientist.

Federal grants are awarded to institutions, not to individuals, so all records and materials developed and all equipment purchased under a grant belong to the grantee institution. Any use of such material by other scientists or by the same scientist who changes employers must be arranged via a *Material Transfer Agreement*.[63] If the samples contain information about human subjects, then the considerations apply that were discussed in chapter 5, in the Research with Human Participants section. Private foundations that support research will have their own rules about ownership of laboratory records.

However, some funding agencies require that data and materials, as well as published papers, developed under their support be made available for sharing with other scientists in order to further scientific cooperation and progress. The 2016 National Science Foundation Award and Administration Guide (current version available on the NSF website), requires that publications from funded research are deposited in a public access compliant repository (as identified in the Public Access Policy); are available for download, reading, and analysis within twelve months of publication; possess a minimum set of machine-readable metadata elements as described in the Public Access Policy; and are reported in annual and final reports with a persistent identifier. Either the final printed version or the final peer-reviewed manuscript is acceptable for deposit.

Furthermore, NSF requires (see website) that "Investigators ... share with other researchers, at no more than incremental cost and within a reasonable time, the primary data, samples, physical collections and other supporting materials created or gathered in the course of work under NSF grants." The other government agencies have similar requirements.

Freedom of Information Act

If your research is funded by the federal government, who can have access to your unpublished notes, data, and special materials, or to your research proposals and grant reports?

The 1966 Freedom of Information Act (FOIA) requires United States government agencies to disclose *agency records* upon request.[64] Anyone can request access to such records by using a form on the FOIA website and paying for the search and copies. These agency records include funded research proposals, progress reports, and final reports. Originally, laboratory records developed under a federal *grant* were not seen as agency records and were not subject to FOIA, but laboratory records developed under a *contract* were subject to FOIA. However, that law was changed by the Shelby Amendment in 1999, and now laboratory records that pertain to research performed after 1999 under federal grants or contracts, that (a) is published and (b) used to affect a federal regulation, are subject to FOIA disclosure. *Records* means written and computer records, but does not include materials or specimens.

There are also nine exemptions from FOIA that an agency may or may not choose to invoke. The FOIA website gives the exemptions as follows:

Exemption 1: Information that is classified to protect national security.
Exemption 2: Information related solely to the internal personnel rules and practices of an agency.
Exemption 3: Information that is prohibited from disclosure by another federal law.

Exemption 4: Trade secrets or commercial or financial information that is confidential or privileged.

Exemption 5: Privileged communications within or between agencies, including:

1. Deliberative Process Privilege
2. Attorney-Work Product Privilege
3. Attorney-Client Privilege

Exemption 6: Information that, if disclosed, would invade another individual's personal privacy.

Exemption 7: Information compiled for law enforcement purposes ...

Exemption 8: Information that concerns the supervision of financial institutions.

Exemption 9: Geological information on wells.

The exemptions most relevant for scientists are numbers 4, 5, and 6.[65] In exemption 4, trade secrets are protected by law, but the contents of *funded* federal grant applications are not considered trade secrets and thus those documents can be released. The contents of *unfunded* federal grant applications are not released, under exemptions 5 (deliberative process) and 6 (personal privacy).

Agencies such as NSF and NIH give detailed FOIA policies on their webpages. These laws and regulations do change, so you are advised to consult the website of the funding agency to be sure that your understanding is current. Your institution will have an intellectual property officer who can help you if you encounter FOIA problems.

Summary: Intellectual Property

Scientists need to understand the basics of intellectual property law, especially with respect to copyrights and patents, in order to protect themselves and their work, no matter whether they work in industry, government, or academia. Working scientists will need the help of a good book (see Further Reading, to follow), will need to search websites for up-to-date information, and will need to consult the intellectual property professionals for any complicated situations. Just as experimental physicists need to know enough about machine shop work to be able to talk to machinists, scientists need to know enough about intellectual property law to be able to talk to intellectual property specialists.

Guided Case Study on Intellectual Property: Confidential Information

You have been an employee of company A for ten years, during which you worked on the development of new biodegradable plastics. You have confidential information (trade secrets) about the products of company A. Now you have moved to company B and your job involves projects for which that confidential information from company A is relevant and very valuable. Can you use that confidential information?

1. Name the values that are involved and consider the conflicts among them. Recall the values of truth and justice from chapter 1. How are those values relevant to this problem?

2. Would more information be helpful? Did you sign a nondisclosure agreement with company A? If you did not, then are you free to use the information? Is the information you have truly proprietary, or could it be considered general technical knowledge? Discussions with your supervisor and with the company B lawyers may be useful.

3. List as many solutions as possible. If the information is deemed not proprietary, then you can proceed to use it. If it is proprietary, then can you try to manage without the information by choosing a different path for your work? Can you seek an agreement with company A to license the use of the information?

4. Do any of these possible solutions require still more information? Can an intellectual property expert help you to determine the level of confidentiality for the information? Can you search the literature to find another solution and to find whether the trade secret is actually not secret? Is company B willing to pay for a license from company A?

5. Think further about how you would implement these solutions. Discussions with your supervisor, an intellectual property expert, and a lawyer are in order. How can you talk with these people without revealing the trade secret?

6. Check on the status of the problem. Has anything changed? Maybe the information has appeared in a publicly published document.

7. Decide on a course of action. To use the information or not to use it is the question, and if using it, how to do so ethically.

Discussion Questions on Intellectual Property

1. What is the difference between a federal *grant* and a federal *contract* (sometimes termed a *procurement contract*)? Which is an academic scientist more likely to encounter? Which is a scientist in industry more likely to encounter?

2. How can the Freedom of Information Act affect a scientist?

3. The website *Sci-Hub* originated in Russia in 2011 and makes available tens of millions of pirated scientific papers obtained by hacking into computer systems.[66] What problems can arise if many scientists use Sci-Hub instead of legitimate access to the literature?

Case Studies on Intellectual Property

1. You are a graduate student in physics and you have collected neutron scattering data on a solid compound that may have interesting and useful electronic properties. Your work was supported by the National Science Foundation and by the neutron facility at the National Institute of Standards and Technology. A graduate student at another university heard you speak about your work at a Gordon Conference and now asks you for a set of your data to study.

    a. Assume that you have not yet written your dissertation or published your analysis of the data. Do you share the data?

    b. Assume that you have submitted your dissertation and published your analysis of the data. Do you share the data?

    c. Under what conditions would a collaboration be in order?

    d. Do you plan to make your full data set available in some way (see chapter 3)?

    e. What if your dissertation advisor disagrees with your decision in each case?

2. You have worked for a company and in the course of your work you have filled ten laboratory notebooks with records of experiments and with ideas for new work. Can you take these notebooks with you if you move to a new job at a different company? At a university?

Inquiry Questions on Intellectual Property

1. You have written and copyrighted a textbook on general chemistry that has become a national bestseller and that has made you a lot of money (more money annually than your university salary). You produced this book partly during your work day, although you did a lot of work during evenings and weekends, and you used your university office, your office computer, and the university library in your writing.

a. Who owns the copyright? You may need to ask someone at your institution.

b. Are you *legally* required to give any of your royalties to the university?

c. Are you *ethically* required to give any of your royalties to the university?

d. Should you use student labor, paid for by the university, to prepare the new edition of your book?

2. You are a faculty member at a large state university. You have developed an online course on research ethics.

a. You developed this course on your own, with no requirement from the university that you do so. Who owns the rights to this course? If you move to another university, can you take the course with you and use it there?

b. You developed this course with the support of a grant from the NSF. Who owns the rights to this course? If you move to another university, can you take the course with you and use it there?

c. You developed this course with the support of a grant from the NSF. A publisher approaches you about making this a commercial online course. Can you legally sell the course to the publisher?

3. You have a CD of the bibliography program EndNote, which costs about $300 to buy. Your colleague asks to borrow the CD to install the program on her computer. Is this illegal? Is this unethical? Read a software end-user's agreement and summarize its points.

4. You have been a faculty member at University X and you did not get tenure, so you are moving to University Y.

a. The equipment in your laboratory was all purchased with a grant from the NIH. Can you take that equipment with you?

b. You have twenty laboratory notebooks and many computer files, all developed while you were employed at University X. Can you take these records to University Y?

c. When you arrive at University Y, you prepare and publish a textbook based on a course the that you developed at University X. Who owns the copyright to the textbook?

5. Suppose you are a graduate student at your university and you have made a discovery that you and your advisor think should be patented. How you would go about doing that within your campus structure?

6. You have published a paper in the American Chemical Society journal *Macromolecules* about the polymerization conditions for compound C. Now you are writing a review paper on polymerization in similar compounds and you want to include a figure from your paper on compound C. Do you need permission from the journal? What if you alter the figure to make it consistent with other figures in your new review article: Do you still need permission?

7. The website *ResearchGate* posts research articles for public use. Can you post your own articles published in journals of the American Chemical Society on ResearchGate? What about your articles in journals of the American Physical Society?

Further Reading on Intellectual Property

Biagioli, Marui, and Peter Galison, eds. *Scientific Authorship: Credit and Intellectual Property in Science.* New York: Routledge, 2002.

Charrow, Robert P. *Law in the Laboratory: A Guide to the Ethics of Federally Funded Science Research.* Chicago: University of Chicago Press, 2010.

Rooksby, Jacob H. *The Branding of the American Mind.* Baltimore, MD: Johns Hopkins University Press, 2016.

Notes

1. Michael Riordan, Lillian Hoddeson, and Adrienne W. Kolb, *Tunnel Visions: The Rise and Fall of the Superconducting Super Collider* (Chicago: University of Chicago Press, 2015).

2. David F. Voss and Daniel E. Koshland Jr., "The Lessons of the Supercollider," *Science* 262, no. 5141 (1993): 1799.

3. Herbert Morawetz, *Polymers: The Origins and Growth of a Science* (New York: John Wiley, 1985).

4. Robert A. Millikan, *Biographical Memoir of Albert Abraham Michelson 1852–1931*, vol. 14, Biographical Memoirs (Washington, DC: National Academy of Sciences, 1938).

5. Donald E. Stokes, *Pasteur's Quadrant: Basic Science and Technological Innovation* (Washington, DC: Brookings Institution Press, 1997).

6. Vannevar Bush, *Science, the Endless Frontier: A Report to the President* (Washington, DC: U.S. Government Printing Office, 1945).

7. Daniel S. Greenberg, *Science, Money, and Politics: Political Triumph and Ethical Erosion* (Chicago: University of Chicago Press, 2001).

8. John R. Steelman, *Science and Public Policy: A Report to the President* (Washington, DC: U.S. Government Printing Office, 1947).

9. Congressional Budget Office, *Federal Support for Research and Development* (Washington, DC: Congress of the United States, 2007).

10. Greenberg, *Science, Money, and Politics.*

11. William H. Press, "What's So Special About Science (and How Much Should We Spend on It?)," *Science* 342, no. 6160 (2013): 817–822.

12. Greenberg, *Science, Money, and Politics.*

13. Jeffrey Mervis, "Building a Scientific Legacy on a Controversial Foundation," *Science* 321, no. 5888 (2008): 480–483.

14. Ibid.

15. Roger A. Pielke and Radford Byerly, "Beyond Basic and Applied," *Physics Today* 51, no. 2 (1998): 42–46.

16. Richard M. Jones, "Senator Barbara Mikulski on 'Scientific Patriotism,'" *FYI: The AIP Bulletin of Science Policy News* 20, February 14, 1994, https://www.aip.org/fyi/1994/senator-barbara-mikulski-scientific-patriotism.

17. Stokes, *Pasteur's Quadrant.*

18. *The National Science Foundation Proposal and Award Policies and Procedures Guide* (Arlington, VA: National Science Foundation, 2014).

19. Steven Weinberg, "The Crisis of Big Science," *New York Review of Books* 59, no. 8 (2012): 59–62.

20. J. Bronowski, *Science and Human Values* (New York: Harper and Row, 1965).

21. Cary Funk and Sara Kehaulani Goo, *A Look at What the Public Does and Does Not Know About Science* (Washington, DC: Pew Research Center, 2015).

22. Pew Research Center, *Coverage Error in Internet Surveys: Who Web-Only Surveys Miss and How That Affects Results* (Washington, DC: Pew Research Center, 2015).

23. National Science Board, *Science and Engineering Indicators 2016* (Washington, DC: National Science Foundation, 2016).

24. Jeffrey Mervis, "For Innovation, No Magic in 3% Rule," *Science* 351, no. 6276 (2016): 900.

25. Peter S. Temes, *The Just War: An American Reflection on the Morality of War in Our Time* (Chicago: Ivan R. Dee, 2003).

26. Jeffrey Kovac, "Science, Ethics and War: A Pacifist's Perspective," *Science and Engineering Ethics* 19, no. 2 (2013): 449–460.

27. Daniel Charles, *Mastermind: The Rise and Fall of Fritz Haber, the Nobel Laureate Who Launched the Age of Chemical Warfare* (New York: HarperCollins, 2005).

28. Vaclav Smil, *Enriching the Earth: Fritz Haber, Carl Bosch, and the Transformation of World Food Production* (Cambridge, MA: MIT Press, 2001).

29. Patrick Coffey, *Cathedrals of Science: The Personalities and Rivalries That Made Modern Chemistry* (New York: Oxford University Press, 2008).

30. Ruth Lewin Sime, *Lise Meitner: A Life in Physics* (Berkeley: University of California Press, 1996).

31. Ronald W. Clark, *Einstein: The Life and Times* (New York: World Publishing, 1971).

32. David C. Cassidy, *Uncertainty: The Life and Science of Werner Heisenberg* (New York: W. H. Freeman, 1992).

33. Klaus Mayer, Maria Wallenius, Klaus Lîtzenkirche, Joan Horta, Adrian Nicholl, Gert Rasmussen, Pieter van Belle et al., "Uranium from German Nuclear Power Projects of the 1940s: A Nuclear Forensic Investigation," *Angewandte Chemie International Edition* 54, no. 45 (2015): 13452–13456.

34. Michael Frayn, *Copenhagen* (London: Methuen Drama, 1998).

35. John Cornwell, *Hitler's Scientists: Science, War, and the Devil's Pact* (New York: Viking, 2003); Thomas Powers, *Heisenberg's War* (New York: Alfred A. Knopf, 1993).

36. Charles, *Mastermind*.

37. Richard Rhodes, *Dark Sun: The Making of the Hydrogen Bomb* (New York: Simon and Schuster, 1996).

38. Jonathan B. Tucker, *War of Nerves: Chemical Warfare from World War I to Al-Qaeda* (New York: Anchor, 2007); Ad Maas and Hans Hooijmaijers, eds., *Scientific Research in World War II: What Scientists Did in the War* (New York: Routledge, 2009).

39. Thomas Hager, *Force of Nature: The Life of Linus Pauling* (New York: Simon and Schuster, 1995).

40. Ann Finkbeiner, *The Jasons: The Secret History of Science's Postwar Elite* (New York: Viking/Penguin, 2006).

41. Annie Jacobsen, *The Pentagon's Brain: An Uncensored History of DARPA, America's Top-Secret Military Research Agency* (New York: Little, Brown, and Company, 2015).

42. Finkbeiner, *The Jasons.*

43. Joseph Borkin, *The Crime and Punishment of I. G. Farben* (New York: Free Press, 1978).

44. Richard M. Price, *The Chemical Weapons Taboo* (Ithaca, NY: Cornell University Press, 1997).

45. "The World Wars," in *The Encyclopedia Britannica*, ed. Robert McHenry (Chicago: University of Chicago Press, 1992).

46. David E. Sanger, "Obama Order Sped up Wave of Cyberattacks against Iran," *New York Times*, June 1, 2012, A1; David E. Sanger, *Confront and Conceal: Obama's Secret Wars and Surprising Use of American Power* (New York: Crown Publishing, 2012).

47. Juliet Nicholson, *The Great Silence: Britain from the Shadow of the First World War to the Dawn of the Jazz Age* (New York: Grove Press, 2009).

48. John Bohannon, "Torture Report Prompts APA Apology," *Science* 349, no. 6245 (2015): 221–222.

49. Robert P. Charrow, *Law in the Laboratory: A Guide to the Ethics of Federally Funded Research* (Chicago: University of Chicago Press, 2010).

50. Ibid.

51. Daniel S. Greenberg, *Science for Sale: The Perils, Rewards, and Delusions of Campus Capitalism* (Chicago: University of Chicago Press, 2007).

52. Nancy Forbes, "Managing Conflicts of Interest," *Industrial Physicist* 7, no. 4 (2001): 22–25.

53. Richard W. Pollay, "Propaganda, Puffing, and the Public Interest," *Public Relations Review* 16, no. 3 (1990): 39–54.

54. Eliot Marshall, "University of Texas Revamps Conflict Rules after Critical Review," *Science* 338, no. 6113 (2012): 1407.

55. Justin Gillis, "Smithsonian Will Tighten Its Guidelines on Disclosure," *New York Times*, June 27, 2015, A10.

56. Caroline Whitbeck, *Ethics in Engineering Practice and Research*, 2nd ed. (Cambridge, UK: Cambridge University Press, 2014).

57. Francis L. Macrina, *Scientific Integrity: Text and Cases in Responsible Conduct of Research*, 4th ed. (Washington, DC: American Society for Microbiology Press, 2014).

58. Charrow, *Law in the Laboratory.*

59. Ibid.

60. Bruce Boyden, *The Failure of the DMCA Notice and Takedown System: A Twentieth Century Solution to a Twenty-First Century Problem* (Fairfax, VA: Center for the Protection of Intellectual Property, George Mason University, 2013).

61. Nick Taylor, *Laser: The Inventor, the Nobel Laureate, and the Thirty-Year Patent War* (New York: Simon and Schuster, 2000).

62. Thomas F. Budinger and Miriam D. Budinger, *Ethics of Emerging Technologies: Scientific Facts and Moral Challenges* (New York: John Wiley and Sons, 2006).

63. Charrow, *Law in the Laboratory*.

64. Ibid.

65. Ibid.

66. John Bohannon, "Who's Downloading Pirated Papers? Everyone," *Science* 352, no. 6285 (2016): 508–512.

Conclusion: Final Advice on Doing Good Science

Two things fill the mind with ever new and increasing admiration and awe, the oftener and the more steadily we reflect on them: the starry heavens above and the moral law within.

—Immanuel Kant, *Critique of Pure Reason*, 1781[1]

Scientists often speak of *doing good science*: *good* in the sense of making a useful contribution to an important question about the world, *good* in the sense of being honest in its development and interpretation of information about the world, and *good* in the sense of being just in the treatment of other people who are a part of scientific enterprise. The study of this and other books on ethics, and the serious contemplation and discussion of ethical quandaries, are a path toward good science. Students who learn to think about ethics in science will then influence their own students and colleagues, and ultimately can affect the very culture of science, bringing science toward a more collegial, cooperative community and toward even better science.

Physical scientists can benefit from other studies to supplement their work in their own specialties and in the ethics of science. You are encouraged to:

• Study the discipline of logic as developed by philosophers, to be better prepared for the possibilities for error in an investigation, either by taking a formal course in logic, or by reading on your own;

• Develop a working understanding of probability, statistics, and design of experiments, the statistical methods for assuring that the results of an experiment are valid;

• Learn more about the history and philosophy of science, and about the lives and careers of other scientists;

• Join the American Association for the Advancement of Science (AAAS) and read the latest articles in the AAAS journal *Science* about matters of ethics and policy in science in the United States and around the world;
• Join the American Chemical Society, the American Physical Society, the Association for Women in Science, or the American Institute of Chemical Engineers, or another society in your discipline, and participate in the societies' conversations about ethics in science.

Doing good science is a noble journey, a journey to "fill the mind with ever new and increasing admiration and awe."[2]

Notes

1. Immanuel Kant, *The Critique of Practical Reason*, 2nd ed., ed. Mortimer J. Adler, trans. Thomas Kingsmill Abbott, vol. 39, Great Books of the Western World (Chicago: Encyclopedia Britannica, 1990), 360.

2. Ibid.

Appendix A: Guidelines for Laboratory Notebooks

Good laboratory recordkeeping matters for efficient data management, for thoughtful analysis of results, for comparisons to later results, and for verifications for patents.[1] Notebooks are appropriate both for laboratory science and for computer or theoretical work. Notebooks can be partly electronic, but paper notebooks make an invaluable permanent diary of the research. These guidelines refer to paper notebooks, but the same general rules apply to electronic records.

1. Use bound notebooks with good-quality paper.
2. Use a different notebook for each project and (usually) for each person.
3. Write in indelible ink. Do not skip pages. Strike through blank parts of pages so that the record is unalterable.
4. Date each page, writing down the year explicitly. You do not want to have to figure out later, for a patent application or for a question about the sequence of events in an experiment, whether 9/25 means September 25, 2016, or September 25, 2017.
5. Maintain a table of contents.
6. Start a project by writing out the nature of and motivation for the investigation. In the course of your work, continue to write down your reasons for each step and your conclusions for each procedure.
7. Write as you work. Do not write things down on scrap paper for later transfer to your notebook. Neatness is not so important that you should jeopardize accuracy.
8. Write into your notebook all the details about your samples: sources, characterization information. It is useful to make a table of literature references and properties for each of your materials. Paste copies of purchase orders into your notebook to simplify reordering.
9. Write into your notebook all the details about your instruments, including manufacturers, model numbers, and calibrations.

10. Write out or paste in copies of all your procedures and protocols. The notebook should be sufficiently complete that another scientist could read it, follow your logic, and duplicate your experiments.

11. Never erase or obscure an error. Simply cross it out and leave it legible, while writing in a correction.

12. When there may be issues of patents and intellectual property, sign and date each page. Periodically have a witness read and sign as well.

13. Keep a backup copy of your notebook. Strictly speaking, laboratory notebooks belong to the institution or company at which the work is done, not to the individual scientist. Thus a second copy should be kept in case the first is lost or damaged, and as a personal copy for the scientist, should he or she change institutions. In some cases, the scientist who leaves an institution or a company will not be allowed to keep copies of notebooks.

Notes

1. Howard M. Karnare, *Writing the Laboratory Notebook* (Washington, DC: The American Chemical Society, 1985); E. Bright Wilson, *An Introduction to Scientific Research* (New York: McGraw-Hill, 1952).

Appendix B: Bibliography

Of making many books there is no end, and much study is a weariness of the flesh.
—Ecclesiastes 12:12, The Holy Bible, King James Version

Books on Ethics in Science

A number of good books on the subject of ethics and science have appeared in the last twenty years.

Briggle, Adam, and Carl Mitcham. *Ethics and Science: An Introduction*. Cambridge, UK: Cambridge University Press, 2012.
Philosophers look at science and the community of scientists.

Committee on Science, Engineering, and Public Policy of the National Academy of Sciences, National Academy of Engineering, and National Academy of Medicine. *On Being a Scientist*. 3rd ed. Washington, DC: National Academy Press, 2009.
This short booklet has been a valuable resource since its first edition in 1989. It is available as a free file from the National Academy Press website.

Elliott, Deni., and Judy E. Stern, eds. *Research Ethics: A Reader*. Hanover, NH: University Press of New England for the Institute for the Study of Applied and Professional Ethics at Dartmouth College, 1997.
Elliott is a philosopher and Stern is in medicine. Their edited book has useful case studies.

Kovac, Jeffrey. *The Ethical Chemist: Professionalism and Ethics in Science*. Upper Saddle River, NJ: Pearson Prentice-Hall, 2004.
Kovac has brief introductions to ethical theory and to professional ethics, followed by fifty case studies for class discussions.

Macrina, Francis L. *Scientific Integrity: Text and Cases in Responsible Conduct of Research*. 4th ed. Washington, DC: American Society for Microbiology Press, 2014.
Macrina's book is aimed at researchers in microbiology and medicine and uses examples from those fields.

Seebauer, Edmund G., and Robert L. Berry. *Fundamentals of Ethics for Scientists and Engineers.* New York: Oxford University Press, 2001.
This book uses the framework of virtue ethics to analyze ethical dilemmas.

Shamoo, Adil E., and David B. Resnik. *Responsible Conduct of Research.* 3rd ed. New York: Oxford University Press, 2015.
Shamoo is a biochemist and Resnik is a philosopher. Their book is aimed at medical researchers.

Shrader-Frechette, Kristin. *Ethics of Scientific Research.* Lanham, MD: Rowman and Littlefield, 1994.
Shrader-Frechette is a philosopher of science and ethics with a strong interest in environmental issues. The chapter "Gender and Racial Biases in Scientific Research" by Helen Longino is unique among these books.

Sigma Xi, The Scientific Research Society. *Honor in Science.* Research Triangle Park, NC: Sigma Xi, 2000.
This brief booklet, available free online, is focused on the basics of research integrity.

Steneck, Nicholas H. *ORI Introduction to the Responsible Conduct of Research.* Washington, DC: Office of Research Integrity, U.S. Department of Health and Human Services, 2007.
Steneck, a historian, is director of the Research Ethics and Integrity Program of the Michigan Institute for Clinical and Health Research. He has written a readable introduction with case studies.

Whitbeck, Caroline. *Ethics in Engineering Practice and Research.* 2nd ed. Cambridge, UK: Cambridge University Press, 2014.
Whitbeck is a philosopher of science and technology, writing for an audience of engineers engaged in both research and practice.

Biographies of Scientists

For scientists who are thinking about how to live their professional lives, the stories of the lives of other scientists and of their handling of ethical issues can be informative and inspirational.

Coffey, Patrick. *Cathedrals of Science: The Personalities and Rivalries That Made Modern Chemistry.* New York: Oxford University Press, 2008.
This book is about several chemists.

Shifman, M. *Physics in a Mad World.* Trans. J. Manteith. Hackensack, NJ: World Scientific, 2015.
This book is about physicists in the 1930s and 1940s.

For individual scientists, see the following books:

John Desmond Bernal (1901–1971)

Brown, Andrew. *J. D. Bernal: The Sage of Science*. New York: Oxford University Press, 2007.

Bernal was a brilliant British scientist who worked on the application of x-ray crystallography to biochemical molecules. He was a scientific advisor to the British government during World War II, and he mentored a generation of crystallographers (see chapter 4).

Elizabeth H. Blackburn (1948–)

Brady, Catherine. *Elizabeth Blackburn and the Story of Telomeres: Deciphering the Ends of DNA*. Cambridge, MA: MIT Press, 2007.

Blackburn has faced the problems of being a woman in science, and has been involved in the policy issues of research in stem cells.

William Henry Bragg (1862–1942) and William Lawrence Bragg (1890–1971)

Glazer, A. M., and Patience Thomson, eds. *Crystal Clear: The Autobiographies of Sir Lawrence and Lady Bragg*. New York: Oxford University Press, 2015.

Jenkin, John. *William and Lawrence Bragg, Father and Son: The Most Extraordinary Collaboration in Science*. New York: Oxford University Press, 2011.

The father and son Braggs developed x-ray crystallography and began a dynasty of crystallographers (see chapter 4).

Marie Sklodowska Curie (1867–1934)

Goldsmith, Barbara. *Obsessive Genius: The Inner World of Marie Curie*. New York: W. W. Norton, 2005.

Quinn, Susan. *Marie Curie: A Life*. New York: Simon and Schuster, 1995.

Curie was a woman working in physics in France in the early part of the twentieth century. She remains the only person to have won Nobel Prizes in two areas of science.

Albert Einstein (1875–1955)

Clark, Ronald W. *Einstein: The Life and Times*. New York: World Publishing, 1971.

Isaacson, Walter. *Einstein: His Life and Universe*. New York: Simon and Schuster, 2009.

Levinson, Thomas. *Einstein in Berlin*. New York: Bantam, 2003.

Einstein was the most eminent physicist of the twentieth century. He confronted the ethical issues of Nazi control of German science and then of dealing with the atomic bomb.

Rosalind E. Franklin (1920–1958)

Maddox, Brenda. *Rosalind Franklin: The Dark Lady of DNA*. New York: HarperCollins, 2003.

Sayre, Anne. *Rosalind Franklin and DNA*. New York: W. W. Norton, 2000.

Franklin worked in post–World War II England as a crystallographer. Her x-ray photographs of DNA led to the understanding of DNA structure. Franklin was not given due credit for her work by other scientists (see James Watson, *The Double Helix* [1968], following).

Mary K. Gaillard (1939–)

Gaillard, Mary K. *A Singularly Unfeminine Profession: One Woman's Journey in Physics*. Hackensack, NJ: World Scientific, 2015.
Gaillard is a contemporary particle physicist who worked in the French scientific world for many years before moving to the University of California, Berkeley.

Fritz Haber (1869–1934)

Charles, Daniel. *Mastermind: The Rise and Fall of Fritz Haber, The Nobel Laureate Who Launched the Age of Chemical Warfare*. New York: HarperCollins, 2005.
Haber was a physical chemist who developed the process for turning nitrogen from the air into fertilizer and explosives. He developed gas warfare for Germany during World War I. When the Nazis came to power, his Jewish background led to his downfall.

Werner Heisenberg (1901–1976)

Cassidy, Daniel C. *Uncertainty: The Life and Science of Werner Heisenberg*. New York: W. H. Freeman, 1992.

Powers, Thomas. *Heisenberg's War: The Secret History of the German Bomb. New York.* Alfred: Knopf, 1993.

Heisenberg, famous for his work in quantum mechanics, worked within the Nazi establishment during World War II, where he was in charge of German research in nuclear physics.

Dorothy Crowfoot Hodgkin (1910–1994)

Ferry, Georgina. *Dorothy Hodgkin: A Life*. Cold Spring Harbor, ME: Cold Spring Harbor Laboratory Press, 2000.
Hodgkin was a British crystallographer who won the Nobel Prize in Chemistry for her elucidation of the structure of vitamin B_{12}. While doing this work, she was raising three children and suffering from rheumatoid arthritis. Later she was a strong advocate for nuclear disarmament.

Hope Jahren (1969–)

Jahren, Hope. *Lab Girl*. New York: Alfred A. Knopf, 2016.

Jahren is a geobiologist whose memoir tells about her growth as a scientist, her love of the laboratory, her experiences of sexism in science, and the relationships that sustain her.

Barbara McClintock (1902–1992)

Keller, Evelyn Fox. *A Feeling for the Organism: The Life and Work of Barbara McClintock*. New York: W. H. Freeman, 1983.

Cytogeneticist McClintock faced the career difficulties of women scientists of her time. Her scientific work first met with hostile resistance from her colleagues, but later was accepted and led to the Nobel Prize in 1983.

Lise Meitner (1878–1968)

Rife, Patricia. *Lise Meitner and the Dawn of the Nuclear Age*. New York: Springer, 1999.

Sime, Ruth Lewin. *Lise Meitner: A Life in Physics*. Berkeley: University of California Press, 1996.

Meitner was a Jewish woman working in Berlin between the world wars. She was a part of the team that discovered nuclear fission, but she was excluded from the Nobel Prize in Chemistry. She declined to work with the allies on the atomic bomb.

J. Robert Oppenheimer (1904–1967)

Bird, Kai, and Martin J. Sherwin. *American Prometheus: The Triumph and Tragedy of J. Robert Oppenheimer*. New York: Alfred A. Knopf, 2006.

Theoretical physicist Oppenheimer led the American effort to build the first atomic bomb. It was he who said that "the physicists have known sin."[1]

Louis Pasteur (1822–1895)

Debré, Patrice. *Louis Pasteur*. Trans. E. Forster. Baltimore, MD: Johns Hopkins University Press, 1998.

Geison, Gerald L. *The Private Science of Louis Pasteur*. Princeton, NJ: Princeton University Press, 1995.

Pasteur was a careful, thoughtful experimentalist who made major contributions to chemistry and to microbiology. He was also an ambitious, solitary, vengeful person who resisted giving credit either to his predecessors or to his collaborators.

Linus Pauling (1901–1994)

Hager, Thomas. *Force of Nature: The Life of Linus Pauling.* New York: Simon and Schuster, 1995.

Pauling was an American chemist who won both the Nobel Prize in Chemistry and the Nobel Peace Prize. He faced such ethical issues as correcting scientific error and the role of scientists in warfare.

Max Planck (1858–1947)

Brown, Brandon R. *Planck: Driven by Vision, Broken by War.* New York: Oxford University Press, 2015.

Planck was a German theoretical physicist who made major contributions to the development of quantum mechanics in 1900–1920. A decent man, he was caught in the web of Nazi evil.

Edward Teller (1908–2003)

Goodschild, Peter. *Edward Teller: The Real Dr. Strangelove.* Cambridge, MA: Harvard University Press, 2004.

Hargittai, Istvan. *Judging Edward Teller: A Closer Look at One of the Most Influential Scientists of the Twentieth Century.* New York: Prometheus, 2010.

Teller, Edward. With Judith L. Shoolery. *Memoirs: A Twentieth-Century Journey in Science and Politics.* Cambridge, MA: Perseus, 2001.

Teller was a physicist born in Hungary and educated in Germany. He emigrated to the United States in 1935. He was involved in the development of the atomic bomb and the hydrogen bomb. He was later a controversial advocate for the Strategic Defense Initiative and other defensive weapons.

Beatrice Hill Tinsley (1941–1981)

Hill, Edward. *My Daughter Beatrice: A Personal Memoir of Dr. Beatrice Tinsley, Astronomer.* New York: American Physical Society, 1986.

Tinsley wrote over one hundred papers in cosmology, while struggling to balance home and work and before succumbing to cancer at the age of forty.

Alan M. Turing (1912–1954)

Hodges, Andrew. *Alan Turing: The Enigma.* Princeton, NJ: Princeton University Press, 2012.

Turing was a British mathematician and cryptographer who deciphered the Nazi secret codes and developed an early computer. He was prosecuted for homosexual acts in 1952 and subjected to chemical castration. He committed suicide in 1954.

Figure B.1
Astronomer Beatrice Hill Tinsley (1941–1981). Courtesy of the Astronomical Society
of the Pacific.

Geerat J. Vermeij (1946–)

Vermeij, Geerat J. *Privileged Hands: A Scientific Life*. New York: Henry Holt, 1996.
Vermeij is a distinguished biologist and paleontologist who is blind.

James D. Watson (1928–)

Watson, James D. *The Double Helix: A Personal Account of the Discovery of the Structure of DNA*. New York: Athenaeum Press, 1968.
This account by Watson is an example of how *not* to treat your colleagues in science,
but is nonetheless a fascinating picture of the process of science.

Fiction about Ethical Issues in Science

Fiction writers lead us into the hearts and minds of scientists who are doing
science and dealing with other scientists.

Djerassi, Carl. *Cantor's Dilemma*. New York: Penguin, 1991.
Djerassi was a fine chemist who developed the birth control pill. He explored in this
novel several issues in ethics in science: mentorship, data management, careful
reasoning.

Goodman, Allegra. *Intuition*. New York: Dial Press, 2006.
Goodman is a distinguished American novelist who imagines the lives, fears, and
motivations of a group of researchers.

Greer, Andrew Sean. *The Path of Minor Planets*. New York: Picador, 2001.
Greer tells of the lives of a group of astronomers who meet every twelve years to study a recurring comet. He depicts the joys and sorrows of science, including an episode of data trimming (omitting data to make findings appear more precise). *The Search* by C. P. Snow (to follow) has an important role in Greer's novel.

Lipsky, Eleazar. *The Scientists*. New York: Appleton-Century-Crofts, 1959.
Lipsky's novel depicts the mentoring relationship and quarrels about credit.

Lewis, Sinclair. *Arrowsmith*. New York: P. F. Collier, 1925.
In one of the best American novels of the early twentieth century, Lewis wrote about a young man's love of medical science and the realities he finds in his career.

Snow, C. P. *The Search*. New York: Charles Scribner's Sons, 1934.
Snow was both a trained scientist and an accomplished writer. He worked in England between the world wars, and it was he who spoke of the "two cultures" of science and humanities.[2] This novel is based on the Bragg research group (see chapter 4 and the books listed earlier about the Braggs). This is the story of a crystallographer whose initial passion turns to disillusion because of the ethics and politics that he encounters. Written in 1934, *The Search* is still a splendid novel about why and how science is done.

Journals Devoted to Ethical Issues in Science

Quotes are from the websites of the journals.

Accountability in Research, Taylor and Francis, Philadelphia, PA.
This is a "peer-reviewed academic journal ... examining systems for ensuring integrity in the conduct of biomedical research."

Science and Engineering Ethics, Springer, Heidelberg, Germany.
"*Science and Engineering Ethics* is a multi-disciplinary journal that explores ethical issues of direct concern to scientists and engineers. Coverage encompasses professional education, standards and ethics in research and practice, extending to the effects of innovation on society at large."

Organizations and Websites Related to Ethical Issues in Science

Websites are indicated by words that will serve for an Internet search (e.g., National Science Foundation) rather than by an Internet address, since online addresses change.

American Association for the Advancement of Science
American Chemical Society

Association for Women in Science (AWIS)

American Institute of Physics

American Physical Society (APS)

COACh (Committee for the Advancement of Women in Chemistry, at the University of Oregon)

Illinois Institute of Technology Center for the Study of Ethics in the Professions

National Academy of Engineering, Center for Engineering Ethics and Society, Online Ethics Center

National Center for Professional and Research Ethics (NCPRE), University of Illinois at Urbana-Champagne

National Organization of Gay and Lesbian Scientists and Technical Professionals (NOGLSTP)

National Organization for the Professional Advancement of Black Chemists and Chemical Engineers (NOBCChE)

National Science Foundation

Office of Research Integrity (ORI), U.S. Department of Health and Human Services

Poynter Center for the Study of Ethics and American Institutions, Indiana University

Sigma Xi Scientific Research Society

Society for the Advancement of Chicanos/Hispanics and Native Americans in Science (SACNAS)

Notes

1. J. Robert Oppenheimer, "Physics in the Contemporary World," Arthur D. Little Memorial Lecture at M.I.T., November 25, 1947.

2. Charles Percy Snow, *The Two Cultures and the Scientific Revolution* (Cambridge, UK: Cambridge University Press, 1959).

Appendix C: Teaching Ethics

Until about 1995, professors of science or engineering rarely addressed matters of research ethics. It was assumed that students would learn about ethical norms for research by osmosis, from being around active scientists and engineers. After a number of troubling cases of research misconduct became public (see chapter 3), funding agencies such as the National Institutes of Health and the National Science Foundation began to recognize the need for more direct and formal training.

The goals in ethics education are to have a student learn to recognize an ethical issue, to analyze solutions to that issue, and to consider the practical challenges and competing interests in the real world. The goal is not to teach rules for behavior, but rather to teach ways of thinking about ethical issues in research. The focus is more on the daily decisions of the professional lives of well-intentioned people, and less on cases of misconduct and fraud.

Modes of Course Delivery

In response to the need for education in ethics and science, a number of colleges and universities developed courses to introduce students to the ethical issues that scientists and engineers confront in their professional lives.[1]

For example, beginning in 1994, such a course was offered at the University of Maryland, College Park: an elective, one-semester, three-credit course, cross-listed between the Department of Chemistry and Biochemistry and the Department of Chemical and Biomolecular Engineering. The syllabus followed the topics in this book. The class was limited to about twenty students and usually included one or two graduate students and one or two postdoctoral associates, with the remainder being senior under-graduates. The postdoctoral associates assisted in teaching the course, to help prepare them to teach the material in their own careers. Assessments of student progress included a judgment by the instructor on the quality of participation in the discussions, written take-home mid-term and final examinations, a term paper, and an oral presentation on the term paper. The written and oral components were opportunities for students to improve their communication skills.

A two-credit version of the course (without the term paper) was offered just for the graduate students. In addition, one-hour introductions to research ethics were presented for new graduate students or as units in other courses. Half-day workshops were made available for various departments and institutes. The development of the courses and workshops was supported by the Ethics Education in Science and Engineering Program of the National Science Foundation in 2006–2008.

Students are hungry for this kind of course and there was always a waiting list to get into it. Graduated students often wrote back after encountering some problem in their work lives, to report how much this training had helped them.

Teaching Strategies

Research and experience in the teaching of research ethics have led to insights into which pedagogical practices are effective.

1. The involvement of science faculty members as the teachers. The very attention given to research ethics says to students that ethical issues matter.[2] It follows that it is more effective to have research ethics taught by the scientists who are the teachers and mentors, and who are themselves struggling with these ethical problems. Philosophers or administrators may be invited to visit the class, to bring their different perspectives to the students, but the science professors themselves make better, more engaged, and more knowledgeable instructors for courses on ethics in science.

2. The active participation of students in class discussions. Professors of science and engineering are more accustomed to teaching by lecturing, as opposed to guiding discussions in which the students do most of the talking. In an ethics course, there are few or no lectures. The teacher's role is to make sure that the students themselves make the important points that the teacher normally would make in a lecture format. Class size is restricted (fifteen to twenty students) in order to include everyone in the discussions. The class meeting is long enough (150 minutes) for discussions to ripen. Students must have read the assigned material before class. The teacher must be prepared to direct the discussion along productive lines, and to make sure that every student is brought into the discussion. One may fear that the students will be reluctant to speak, but this is not the case. The students are very interested in this material, and they quickly engage with the instructor and their peers.

While individuals at different levels of their careers—undergraduate students, graduate students, postdoctoral researchers—have different needs and experiences, it is still possible to teach all these levels in the same course, at the same time. In fact, the diversity of experiences and outlooks enriches the discussion and the students learn from one another. Cultural diversity can also deepen the discussion. For example, in China, the concept of plagiarism is quite different from how plagiarism is perceived in western culture.[3] The awareness of such differences will give students valuable perspectives.

3. The use of discussion questions, case studies, and role-playing exercises that generate complex ethical scenarios. Questions, case studies, and role-playing exercises need to be complex enough to generate analysis and discussion.[4] Some examples may seem quite easy, but a simple scenario can be made more interesting and challenging by adding complicating factors. Often the instructor and the students can bring to the class the problems with which they are struggling at the time. Sometimes the discussions involve controversial subjects, such as affirmative action. Then there are opportunities for students (and instructors) to exercise skills in respecting opinions that differ from their own.

Further Reading

Pimple, Kevin D. "General Issues in Teaching Research Ethics." In *Research Ethics: Cases and Materials*, ed. Robin Levin Penslaar, 3–12. Bloomington: Indiana University Press, 1995.

Swazey, Judith P., and Stephanie J. Bird. "Teaching and Learning Research Ethics." *Professional Ethics* 4, no. 3–4 (1996): 155–178.

Notes

1. Linda M. Sweeting, "Ethics in Science for Undergraduate Students," *Journal of Chemical Education* 76, no. 3 (1999): 369–372; Brian P. Coppola, "Targeting Entry Points for Ethics in Chemistry Teaching and Learning," *Journal of Chemical Education* 77, no. 11 (2000): 1506–1511; Patricia Ann Maybrook, "Research Skills and Ethics: A Graduate Course Empowering Graduate Students for Productive Research Careers in Graduate School and Beyond," *Journal of Chemical Education* 78, no. 12 (2001): 1628–1631; J. Howard Rytting and Richard L. Schowen, "Issues in Scientific Integrity: A Practical Course for Graduate Students in the Chemical Sciences," *Journal of Chemical Education* 75, no. 10 (1998): 1317–1320; Arri Eisen and Kathy P. Parker, "A Model for Teaching Research Ethics," *Science and Engineering Ethics* 10, no. 4 (2004): 693–704; Judy E. Stern and Deni Elliott, *The Ethics of Scientific Research: A Guidebook for Course Development* (Hanover, NH: University Press of New England, 1997); Jeffrey Kovac, "Scientific Ethics in Chemical Education," *Journal of Chemical Education* 73, no. 11 (1996): 926–928.

2. Kenneth D. Pimple, "General Issues in Teaching Research Ethics," in *Research Ethics: Cases and Materials*, ed. Robin Levin Penslar (Bloomington: Indiana University Press, 1995).

3. P. J. Langlais, "Ethics for the New Generation," *Chronicle of Higher Education* 52, no. 19 (2006): B11.

4. Ralph L. Rosnow, "Teaching Research Ethics through Role-Playing and Discussion," *Teaching of Psychology* 17, no. 3 (1990): 179–181; David B. Strohmetz and Anne A. Skleder, "The Use of Role-Playing in Teaching Research Ethics: A Validation Study," *Teaching of Psychology* 19, no. 2 (1992): 106–108; Patrick E. Hoggard, "Trying a Case on Ethics in Scientific Research: A Role-Playing Exercise for Students and Faculty in a Summer Undergraduate Research Program," *Journal of Chemical Education* 85, no. 6 (2008): 802–804.

Appendix D: Bioethics

Elements of Ethics for Physical Scientists is aimed at chemists and physicists, who are less likely than biologists or medical researchers to encounter the challenges of bioethics. In case these issues do arise, here are selected references for the major topics in bioethics.

General References on Bioethics

Budinger, Thomas F., and Miriam D. Budinger. *Ethics of Emerging Technologies: Scientific Facts and Moral Challenges*. New York: John Wiley, 2006.

Vaughn, Lewis. *Bioethics: Principles, Issues and Cases*. 2nd ed. New York: Oxford University Press, 2012.

Genetic Technology

Genetic screening, cloning, genetic therapies, human genome project.

Barash, Carol Isaacson. *Just Genes: The Ethics of Genetic Technologies*. Santa Barbara, CA: Praeger, 2007.

Assisted Reproductive Technology

In vitro fertilization, cloning, surrogacy, fertility treatments, artificial insemination, egg donation.

Hull, Richard T. *Ethical Issues in the New Reproductive Technologies*. 2nd ed. Amherst, NY: Prometheus Books, 2005.

Stem Cell Technology

Hyun, Insoo. *Bioethics and the Future of Stem Cell Research*. Cambridge, UK: Cambridge University Press, 2013.

Ruse, Michael, and Christopher A. Pynes, eds. *The Stem Cell Controversy: Debating the Issues*. 2nd ed. Amherst, NY: Prometheus, 2006.

Index